VLSI Physical Design:
From Graph Partitioning to Timing Closure

Andrew B. Kahng • Jens Lienig
Igor L. Markov • Jin Hu

VLSI Physical Design:
From Graph Partitioning
to Timing Closure

 Springer

Andrew B. Kahng
University of California at San Diego
Departments of CSE and ECE
Mailcode #0404
La Jolla, California 92093
USA
abk@ucsd.edu

Jens Lienig
Dresden University of Technology
Electrical Engineering and
Information Technology
Helmholtzstr. 10
01069 Dresden
Germany
jens@ieee.org

Igor L. Markov
University of Michigan
Electrical Engineering and
Computer Science
2260 Hayward St.
Ann Arbor, Michigan 48109
USA
imarkov@eecs.umich.edu

Jin Hu
University of Michigan
Electrical Engineering and
Computer Science
2260 Hayward St.
Ann Arbor, Michigan 48109
USA
jinhu@eecs.umich.edu

ISBN 978-94-007-9020-9   ISBN 978-90-481-9591-6 (eBook)
DOI 10.1007/978-90-481-9591-6
Springer Dordrecht Heidelberg London New York

Cover design: eStudio Calamar S.L.

Printed on acid-free paper

Springer is part of Springer Science+Business Media (www.springer.com)

Foreword

Physical design of integrated circuits remains one of the most interesting and challenging arenas in the field of Electronic Design Automation. The ability to integrate more and more devices on our silicon chips requires the algorithms to continuously scale up. Nowadays we can integrate 2e9 transistors on a single 45nm-technology chip. This number will continue to scale for the next couple of technology generations, requiring more transistors to be automatically placed on a chip and connected together. In addition, more and more of the delay is contributed by the wires that interconnect the devices on the chip. This has a profound effect on how physical design flows need to be put together. In the 1990s, it was safe to assume that timing goals of the design could be reached once the devices were placed well on the chip. Today, one does not know whether the timing constraints can be satisfied until the final routing has completed.

As far back as 10 or 15 years ago, people believed that most physical design problems had been solved. But, the continued increase in the number of transistors on the chip, as well as the increased coupling between the physical, timing and logic domains warrant a fresh look at the basic algorithmic foundations of chip implementation. That is exactly what this book provides. It covers the basic algorithms underlying all physical design steps and also shows how they are applied to current instances of the design problems. For example, Chapter 7 provides a great deal of information on special types of routing for specific design situations.

Several other books provide in-depth descriptions of core physical design algorithms and the underlying mathematics, but this book goes a step further. The authors very much realize that the era of individual point algorithms with single objectives is over. Throughout the book they emphasize the multi-objective nature of modern design problems and they bring all the pieces of a physical design flow together in Chapter 8. A complete flow chart, from design partitioning and floorplanning all the way to electrical rule checking, describes all phases of the modern chip implementation flow. Each step is described in the context of the overall flow with references to the preceding chapters for the details.

This book will be appreciated by students and professionals alike. It starts from the basics and provides sufficient background material to get the reader up to speed on the real issues. Each of the chapters by itself provides sufficient introduction and depth to be very valuable. This is especially important in the present era, where experts in one area must understand the effects of their algorithms on the remainder of the design flow. An expert in routing will derive great benefit from reading the chapters on planning and placement. An expert in Design For Manufacturability (DFM) who seeks a better understanding of routing algorithms, and of how these algorithms can be affected by choices made in setting DFM requirements, will benefit tremendously from the chapters on global and detailed routing.

The book is completed by a detailed set of solutions to the exercises that accompany each chapter. The exercises force the student to truly understand the basic physical design algorithms and apply them to small but insightful problem instances.

This book will serve the EDA and design community well. It will be a foundational text and reference for the next generation of professionals who will be called on to continue the advancement of our chip design tools.

Dr. Leon Stok
Vice President, Electronic Design Automation
IBM Systems and Technology Group
Hopewell Junction, NY

Preface

VLSI physical design of integrated circuits underwent explosive development in the 1980s and 1990s. Many basic techniques were suggested by researchers and implemented in commercial tools, but only described in brief conference publications geared for experts in the field. In the 2000s, academic and industry researchers focused on comparative evaluation of basic techniques, their extension to large-scale optimization, and the assembly of point optimizations into multi-objective design flows. Our book covers these aspects of physical design in a consistent way, starting with basic concepts in Chapter 1 and gradually increasing the depth to reach advanced concepts, such as physical synthesis. Readers seeking additional details, will find a number of references discussed in each chapter, including specialized monographs and recent conference publications.

Chapter 2 covers netlist partitioning. It first discusses typical problem formulations and proceeds to classic algorithms for balanced graph and hypergraph partitioning. The last section covers an important application – system partitioning among multiple FPGAs, used in the context of high-speed emulation in functional validation.

Chapter 3 is dedicated to chip planning, which includes floorplanning, power-ground planning and I/O assignment. A broad range of topics and techniques are covered, ranging from graph-theoretical aspects of block-packing to optimization by simulated annealing and package-aware I/O planning.

Chapter 4 addresses VLSI placement and covers a number of practical problem formulations. It distinguishes between global and detailed placement, and first covers several algorithmic frameworks traditionally used for global placement. Detailed placement algorithms are covered in a separate section. Current state of the art in placement is reviewed, with suggestions to readers who might want to implement their own software tools for large-scale placement.

Chapters 5 and 6 discuss global and detailed routing, which have received significant attention in research literature due to their interaction with manufacturability and chip-yield optimizations. Topics covered include representing layout with graph models and performing routing, for single and multiple nets, in these models. State-of-the-art global routers are discussed, as well as yield optimizations performed in detailed routing to address specific types of manufacturing faults.

Chapter 7 deals with several specialized types of routing which do not conform with the global-detailed paradigm followed by Chapters 5 and 6. These include non-Manhattan area routing, commonly used in PCBs, and clock-tree routing required for every synchronous digital circuit. In addition to algorithmic aspects, we explore the impact of process variability on clock-tree routing and means of decreasing this impact.

Chapter 8 focuses on timing closure, and its perspective is particularly unique. It offers a comprehensive coverage of timing analysis and relevant optimizations in placement, routing and netlist restructuring. Section 8.6 assembles all these techniques, along with those covered in earlier chapters, into an extensive design flow, illustrated in detail with a flow chart and discussed step-by-step with several figures and many references.

This book does not assume prior exposure to physical design or other areas of EDA. It introduces the reader to the EDA industry and basic EDA concepts, covers key graph concepts and algorithm analysis, carefully defines terms and specifies basic algorithms with pseudocode. Many illustrations are given throughout the book, and every chapter includes a set of exercises, solutions to which are given in one of the appendices. Unlike most other sources on physical design, we made an effort to avoid impractical and unnecessarily complicated algorithms. In many cases we offer comparisons between several leading algorithmic techniques and refer the reader to publications with additional empirical results.

Some chapters are based on material in the book *Layoutsynthese elektronischer Schaltungen – Grundlegende Algorithmen für die Entwurfsautomatisierung*, which was published by Springer in 2006.

We are grateful to our colleagues and students who proofread earlier versions of this book and suggested a number of improvements (in alphabetical order): Matthew Guthaus, Kwangok Jeong, Johann Knechtel, Andreas Krinke, Nancy MacDonald, Jarrod Roy, Yen-Kuan Wu and Hailong Yao.

Images for global placement and clock routing in Chapter 8 were provided by Myung-Chul Kim and Dong-Jin Lee. Cell libraries in Appendix B were provided by Bob Bullock, Dan Clein and Bill Lye from PMC Sierra; the layout and schematics in Appendix B were generated by Matthias Thiele. The work on this book was partially supported by the National Science Foundation (NSF) through the CAREER award 0448189 as well as by Texas Instruments and Sun Microsystems.

We hope that you will find the book interesting to read and useful in your professional endeavors.

Sincerely,

Andrew, Jens, Igor and Jin

Table of Contents

xi

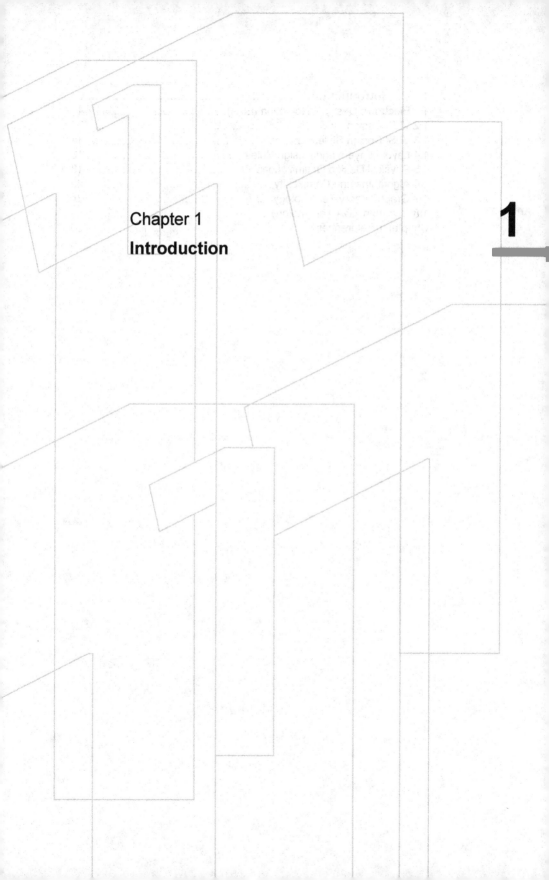

Chapter 1
Introduction

1

1

1 Introduction

The design and optimization of *integrated circuits* (*ICs*) are essential to the production of new semiconductor chips. Modern chip design has become so complex that it is largely performed by specialized software, which is frequently updated to reflect improvements in semiconductor technologies and increasing design complexities. A *user* of this software needs a high-level understanding of the implemented algorithms. On the other hand, a *developer* of this software must have a strong computer-science background, including a keen understanding of how various algorithms operate and interact, and what their performance bottlenecks are.

This book introduces and evaluates algorithms used during *physical design* to produce a geometric chip layout from an abstract circuit design. Rather than list every relevant technique, however, it presents the essential and fundamental algorithms used within each physical design stage.

- *Partitioning* (Chap. 2) and *chip planning* (Chap. 3) of design functionality during the initial stages of physical design
- Geometric *placement* (Chap. 4) and *routing* (Chaps. 5-6) of circuit components
- Specialized routing and *clock tree synthesis* for synchronous circuits (Chap. 7)
- Meeting specific technology and performance requirements, i.e., *timing closure*, such that the final fabricated layout satisfies system objectives (Chap. 8)

Other design steps, such as *circuit design, logic synthesis, transistor-level layout* and *verification*, are not discussed in detail, but are covered in such references as [1.1].

This book emphasizes *digital* circuit design for *very large-scale integration* (*VLSI*); the degree of automation for digital circuits is significantly higher than for *analog* circuits. In particular, the focus is on algorithms for digital ICs, such as system partitioning for *field-programmable gate arrays* (*FPGAs*) or *clock network synthesis* for *application-specific integrated circuits* (*ASICs*). Similar design techniques can be applied to other implementation contexts such as *multi-chip modules* (*MCMs*) and *printed circuit boards* (*PCBs*).

The following broad questions, of interest to both students and designers, are addressed in the upcoming chapters.

- How is functionally correct layout produced from a netlist?
- How does software for VLSI physical design work?
- How do we develop and improve software for VLSI physical design?

More information about this book is at http://vlsicad.eecs.umich.edu/KLMH/.

1.1 Electronic Design Automation (EDA)

The *Electronic Design Automation* (*EDA*) industry develops software to support engineers in the creation of new integrated-circuit (IC) designs. Due to the high complexity of modern designs, EDA touches almost every aspect of the IC design flow, from high-level system design to fabrication. EDA addresses designers' needs at multiple levels of electronic system hierarchy, including integrated circuits (ICs), multi-chip modules (MCMs), and printed circuit boards (PCBs).

Progress in semiconductor technology, based on *Moore's Law* (Fig. 1.1), has led to integrated circuits (1) comprised of hundreds of millions of transistors, (2) assembled into packages, each having multiple chips and thousands of pins, and (3) mounted onto *high-density interconnect* (*HDI*) circuit boards with dozens of wiring layers. This design process is highly complex and heavily depends on automated tools. That is, computer software is used to mostly automate design steps such as logic design, simulation, physical design, and verification.

EDA was first used in the 1960s in the form of simple programs to automate placement of a very small number of blocks on a circuit board. Over the next few years, the advent of the integrated circuit created a need for software that could reduce the total number of gates. Current software tools must additionally consider electrical effects such as signal delays and capacitive coupling between adjacent wires. In the modern VLSI design flow, nearly all steps use software to automate optimizations.

In the 1970s, semiconductor companies developed in-house EDA software, specialized programs to address their proprietary design styles. In the 1980s and 1990s, independent software vendors created new tools for more widespread use. This gave rise to an independent EDA industry, which now enjoys annual revenues of approximately five billion dollars and employs around twenty thousand people. Many EDA companies have headquarters in Santa Clara county, in the state of California. This area has been aptly dubbed the *Silicon Valley*.

Several annual conferences showcase the progress of the EDA industry and academia. The most notable one is the *Design Automation Conference* (*DAC*), which holds an industry trade show as well as an academic symposium. The *International Conference on Computer-Aided Design* (*ICCAD*) places emphasis on academic research, with papers that relate to specialized algorithm development. PCB developers attend *PCB Design Conference West* in September. Overseas, Europe and Asia host the *Design, Automation and Test in Europe* (*DATE*) conference and the *Asia and South Pacific Design Automation Conference* (*ASP-DAC*), respectively. The world-wide engineering association *Institute of Electrical and Electronic Engineers* (*IEEE*) publishes the monthly *IEEE Transactions on Computer-Aided Design of Integrated Circuits and Systems* (*TCAD*), while the *Association for Computing Machinery* (*ACM*) publishes *ACM Transactions on Design Automation of Electronic Systems* (*TODAES*).

Impact of EDA. According to Moore's Law (Fig. 1.1), the number of transistors on a chip is increasing at an exponential rate. Historically, this corresponds to an annual *compounded* increase of 58% in the number of transistors per chip.

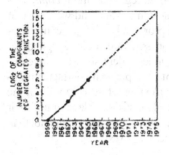

Fig. 1.1 Moore's Law and the original graph from Gordon Moore's article [1.10] predicting the increase in number of transistors. In 1965, Gordon Moore (Fairchild) stated that the number of transistors on an IC would double every year. Ten years later, he revised his statement, asserting that doubling would occur every 18 months. Since then, this "rule" has been famously known as Moore's Law.

However, chip designs produced by prominent semiconductor companies suggest a different trend. The annual *productivity*, measured by the number of transistors, of designers, and (fixed-size) design teams has an annual compounded growth of only around 21% per year, leading to a *design productivity gap* [1.5]. Since the number of transistors is highly context-specific – analog versus digital or memory versus logic – this statistic, due to SEMATECH in the mid-1990s, refers to the design productivity for *standardized transistors*.

Fig 1.2, reproduced from the *International Technology Roadmap for Semiconductors (ITRS)* [1.5], demonstrates that cost-feasible IC products require innovation in EDA technology. Given the availability of efficient design technologies to semiconductor design teams, the hardware design cost of a typical portable system-on-chip, e.g., a baseband processor chip for a cell phone, remains manageable at $15.7 million (2009 estimate). With associated software design costs, the overall chip design project cost is $45.3 million. Without the design technology innovations between 1993 and 2007, and the resulting design productivity improvements, the design cost of a chip would have been $1,800 million, well over a *billion* dollars.

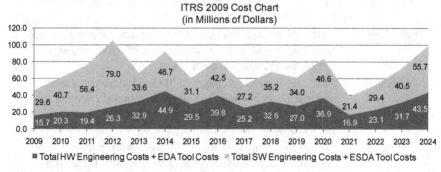

Fig. 1.2 Recent editions of the semiconductor technology roadmap project total hardware (HW) engineering costs + EDA tool costs (dark gray) and total software (SW) engineering costs + electronic system design automation (ESDA) tool costs (light gray). This shows the impact of EDA technologies on overall IC design productivity and hence IC design cost [1.5].

History of EDA. After tools for schematic entry of integrated circuits were developed, the first EDA tool, a placer that optimized the physical locations of devices on a circuit board, was created in the late 1960s. Shortly thereafter, programs were written to aid circuit layout and visualization. The first integrated-circuit *computer-aided design* (*CAD*) systems addressed the physical design process and were written in the 1970s. During that era, most CAD tools were proprietary – major companies such as IBM and AT&T Bell Laboratories relied on software tools designed for internal use only. However, beginning in the 1980s, independent software developers started to write tools that could serve the needs of multiple semiconductor product companies. The electronic design automation (EDA) market grew rapidly in the 1990s, as many design teams adopted commercial tools instead of developing their own in-house versions. The largest EDA software vendors today are, in alphabetical order, *Cadence Design Systems*, *Mentor Graphics*, and *Synopsys*.

EDA tools have always been geared toward automating the entire design process and linking the design steps into a complete design flow. However, such integration is challenging, since some design steps need additional degrees of freedom, and scalability requires tackling some steps independently. On the other hand, the continued decrease of transistor and wire dimensions has blurred the boundaries and abstractions that separate successive design steps – physical effects such as signal delays and coupling capacitances need to be accurately accounted for earlier in the design cycle. Thus, the design process is moving from a sequence of atomic (independent) steps toward a deeper level of integration. Tab. 1.1 summarizes a timeline of key developments in circuit and physical design.

Tab. 1.1 Timeline of EDA progress with respect to circuit and physical design.

Time Period	Circuit and Physical Design Process Advancements
1950-1965	Manual design only.
1965-1975	Layout editors, e.g., place and route tools, first developed for printed circuit boards.
1975-1985	More advanced tools for ICs and PCBs, with more sophisticated algorithms.
1985-1990	First performance-driven tools and parallel optimization algorithms for layout; better understanding of underlying theory (graph theory, solution complexity, etc.).
1990-2000	First over-the-cell routing, first 3D and multilayer placement and routing techniques developed. Automated circuit synthesis and routability-oriented design become dominant. Start of parallelizing workloads. Emergence of physical synthesis.
2000-now	Design for Manufacturability (DFM), optical proximity correction (OPC), and other techniques emerge at the design-manufacturing interface. Increased reusability of blocks, including intellectual property (IP) blocks.

1.2 VLSI Design Flow

The process of designing a very large-scale integrated (VLSI) circuit is highly complex. It can be separated into distinct steps (Fig. 1.3). Earlier steps are high-level; later design steps are at lower levels of abstraction. At the end of the process, before fabrication, tools and algorithms operate on detailed information about each circuit element's geometric shape and electrical properties.

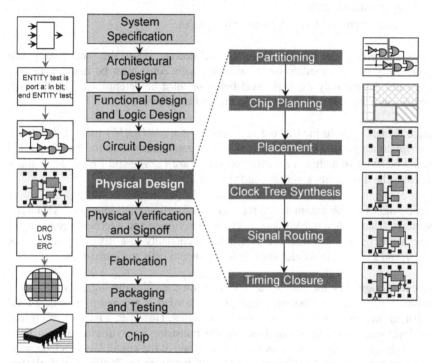

Fig. 1.3 The major steps in the VLSI circuit design flow, with a focus on the physical design steps: partitioning (Chap. 2), chip planning (Chap. 3), placement (Chap. 4), clock tree synthesis (Chap. 7), routing (Chaps. 5-6), and timing closure (Chap. 8).

The steps of the VLSI design flow in Fig. 1.3 are discussed in detail below. For further reading on physical design algorithms, see [1.1]. Books [1.6], [1.9], [1.11] and [1.12] cover other specialized topics that are not addressed here.

System specification. Chip architects, circuit designers, product marketers, operations managers, and layout and library designers collectively define the overall goals and high-level requirements of the system. These goals and requirements span functionality, performance, physical dimensions and production technology.

Architectural design. A basic architecture must be determined to meet the system specifications. Example decisions are

- Integration of analog and mixed-signal blocks
- Memory management – serial or parallel – and the addressing scheme
- Number and types of computational cores, such as processors and digital signal processing (DSP) units – and particular DSP algorithms
- Internal and external communication, support for standard protocols, etc.
- Usage of hard and soft intellectual-property (IP) blocks
- Pinout, packaging, and the die-package interface
- Power requirements
- Choice of process technology and layer stacks

Functional and logic design. Once the architecture is set, the functionality and connectivity of each module (such as a processor core) must be defined. During functional design, only the high-level behavior must be determined. That is, each module has a set of inputs, outputs, and timing behavior.

Logic design is performed at the *register-transfer level* (*RTL*) using a *hardware description language* (*HDL*) by means of programs that define the functional and timing behavior of a chip. Two common HDLs are *Verilog* and *VHDL*. HDL modules must be thoroughly simulated and verified.

Logic synthesis tools automate the process of converting HDL into low-level circuit elements. That is, given a Verilog or VHDL description and a technology library, a logic synthesis tool can map the described functionality to a list of signal nets, or *netlist*, and specific circuit elements such as standard cells and transistors.

Circuit design. For the bulk of digital logic on the chip, the logic synthesis tool automatically converts Boolean expressions into what is referred to as a gate-level netlist, at the granularity of standard cells or higher. However, a number of critical, low-level elements must be designed at the transistor level; this is referred to as *circuit design*. Example elements that are designed at the circuit level include static RAM blocks, I/O, analog circuits, high-speed functions (multipliers), and electrostatic discharge (ESD) protection circuits. The correctness of circuit-level design is predominantly verified by circuit simulation tools such as SPICE.

Physical design. During physical design, all design components are instantiated with their geometric representations. In other words, all macros, cells, gates, transistors, etc., with fixed shapes and sizes per fabrication layer are assigned spatial locations (placement) and have appropriate routing connections (routing) completed in metal layers. The result of physical design is a set of manufacturing specifications that must subsequently be verified.

Physical design is performed with respect to design rules that represent the physical limitations of the fabrication medium. For instance, all wires must be a prescribed minimum distance apart and have prescribed minimum width. As such, the design layout must be recreated in (*migrated* to) each new manufacturing technology.

Physical design directly impacts circuit performance, area, reliability, power, and manufacturing yield. Examples of these impacts are discussed below.

–   Performance: long routes have significantly longer signal delays.
–   Area: placing connected modules far apart results in larger and slower chips.
–   Reliability: large number of vias can significantly reduce the reliability of the circuit.
–   Power: transistors with smaller gate lengths achieve greater switching speeds at the cost of higher leakage current and manufacturing variability; larger transistors and longer wires result in greater dynamic power dissipation.
–   Yield: wires routed too close together may decrease yield due to *electrical shorts* occurring during manufacturing, but spreading gates too far apart may also undermine yield due to longer wires and a higher probability of *opens*.

Due to its high complexity, physical design is split into several key steps (Fig. 1.3).

–   *Partitioning* (Chap. 2) breaks up a circuit into smaller subcircuits or modules, which can each be designed or analyzed individually.
–   *Floorplanning* (Chap. 3) determines the shapes and arrangement of subcircuits or modules, as well as the locations of external ports and IP or macro blocks.
–   *Power and ground routing* (Chap. 3), often intrinsic to floorplanning, distributes power (*VDD*) and ground (*GND*) nets throughout the chip.
–   *Placement* (Chap. 4) finds the spatial locations of all cells within each block.
–   *Clock network synthesis* (Chap. 7) determines the buffering, gating (e.g., for power management) and routing of the clock signal to meet prescribed skew and delay requirements.
–   *Global routing* (Chap. 5) allocates routing resources that are used for connections; example resources include routing tracks in channels and in switchboxes.
–   *Detailed routing* (Chap. 6) assigns routes to specific metal layers and routing tracks within the global routing resources.
–   *Timing closure* (Chap. 8) optimizes circuit performance by specialized placement and routing techniques.

After detailed routing, *electrically-accurate layout optimization* is performed at a small scale. Parasitic resistances (R), capacitances (C) and inductances (L) are extracted from the completed layout, and then passed to timing analysis tools to check the functional behavior of the chip. If the analyses reveal erroneous behavior or an insufficient design margin (*guardband*) against possible manufacturing and environmental variations, then incremental design optimizations are performed.

The physical design of analog circuits deviates from the above methodology, which is geared primarily toward digital circuits. For analog physical design, the geometric representation of a circuit element is created using *layout generators* or manual drawing. These generators only use circuit elements with known electrical parameters, such as the resistance of a resistor, and accordingly generate the appropriate geometric representation, e.g., a resistor layout with specified length and width.

Physical verification. After physical design is completed, the layout must be fully verified to ensure correct electrical and logical functionality. Some problems found during physical verification can be tolerated if their impact on chip yield is negligible. In other cases, the layout must be changed, but these changes must be minimal and should not introduce new problems. Therefore, at this stage, layout changes are usually performed manually by experienced design engineers.

– *Design rule checking* (*DRC*) verifies that the layout meets all technology-imposed constraints. DRC also verifies layer density for *chemical-mechanical polishing* (*CMP*).
– *Layout vs. schematic* (*LVS*) checking verifies the functionality of the design. From the layout, a netlist is derived and compared with the original netlist produced from logic synthesis or circuit design.
– *Parasitic extraction* derives electrical parameters of the layout elements from their geometric representations; with the netlist, these are used to verify the electrical characteristics of the circuit.
– *Antenna rule checking* seeks to prevent *antenna effects*, which may damage transistor gates during manufacturing plasma-etch steps through accumulation of excess charge on metal wires that are not connected to PN-junction nodes.
– *Electrical rule checking* (*ERC*) verifies the correctness of power and ground connections, and that signal transition times (*slew*), capacitive loads and fan-outs are appropriately bounded.

Both analysis and synthesis techniques are integral to the design of VLSI circuits. Analysis typically entails the modeling of circuit parameters and signal transitions, and often involves the solution of various equations using established numerical methods. The choice of algorithms for these tasks is relatively straightforward, compared to the vast possibilities for syntheses and optimization. Therefore, this book focuses on optimization algorithms used in IC physical design, and does not cover computational techniques used during physical verification and signoff.

Fabrication. The final DRC-/LVS-/ERC-clean layout, usually represented in the *GDSII Stream* format, is sent for manufacturing at a dedicated silicon foundry (*fab*). The handoff of the design to the manufacturing process is called *tapeout*, even though data transmission from the design team to the silicon fab no longer relies on magnetic tape [1.6]. Generation of the data for manufacturing is sometimes referred to as *streaming out*, reflecting the use of GDSII Stream.

At the fab, the design is patterned onto different layers using photolithographic processes. Photomasks are used so that only certain patterns of silicon, specified by the layout, are exposed to a laser light source. Producing an IC requires many masks; modifying the design requires changes to some or all of the masks.

ICs are manufactured on round silicon wafers with diameters ranging from 200 mm (8 inches) to 300 mm (12 inches). The ICs must then be tested and labeled as either *functional* or *defective*, sometimes according to *bins* depending on the functional or parametric (speed, power) tests that have failed. At the end of the manufacturing process, the ICs are separated, or *diced*, by sawing the wafer into smaller pieces.

Packaging and testing. After dicing, functional chips are typically packaged. Packaging is configured early in the design process, and reflects the application along with cost and form factor requirements. Package types include *dual in-line packages* (*DIPs*), *pin grid arrays* (*PGAs*), and *ball grid arrays* (*BGAs*). After a die is positioned in the package cavity, its pins are connected to the package's pins, e.g., with *wire bonding* or solder bumps (*flip-chip*). The package is then sealed.

Manufacturing, assembly and testing can be sequenced in different ways. For example, in the increasingly important *wafer-level chip-scale packaging* (*WLCSP*) methodology, "bumping" with high-density solder bumps that facilitate delivery of power, ground and signals from the package to the die is performed before the wafer is diced. With multi-chip module based integration, chips are usually not packaged individually; rather, they are integrated as bare dice into the MCM, which is packaged separately at a later point. After packaging, the finished product may be tested to ensure that it meets design requirements such as function (input/output relations), timing or power dissipation.

1.3 VLSI Design Styles

Selecting an appropriate circuit-design style is very important because this choice affects time-to-market and design cost. VLSI design styles fall in two categories – *full-custom* and *semi-custom*. Full-custom design is primarily seen with extremely high-volume parts such as *microprocessors* or FPGAs, where the high cost of design effort is amortized over large production volumes. Semi-custom design is used more frequently because it reduces the complexity of the design process, and hence time-to-market and overall cost as well.

The following semi-custom standard design styles are the most commonly used.

— *Cell-based*: typically using standard cells and macro cells, the design has many pre-designed elements such as logic gates that are copied from libraries.
— *Array-based*: typically either gate arrays or FPGAs, the design has a portion of pre-fabricated elements connected by pre-routed wires.

Full-custom design. Among available design styles, a full-custom design style has the fewest constraints during layout generation, e.g., blocks can be placed anywhere on the chip without restriction. This approach usually results in a very compact chip with highly optimized electrical properties. However, such effort is laborious, time-consuming, and can be error-prone due to a relative lack of automation.

Full-custom design is primarily useful for microprocessors and FPGAs, where the high cost of design effort is amortized over large production volumes, as well as for analog circuits, where extreme care must be taken to achieve matched layout and adherence to stringent electrical performance specifications.

An essential tool for full-custom design is an efficient layout editor that does more than just draw polygons (Fig. 1.4). Many improved layout editors have integrated DRC checkers that continuously verify the current layout. If all design-rule violations are fixed as they occur, the final layout will be DRC-clean by construction.

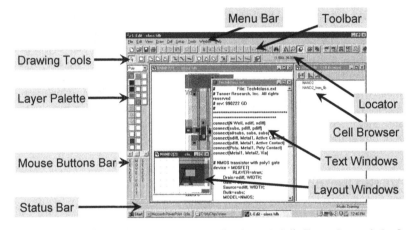

Fig. 1.4 An example of a high-functionality layout editor [From L-Edit, Tanner Research, Inc.].

Standard-cell designs. A digital standard cell is a predefined block that has fixed size and functionality. For instance, an AND cell with two inputs contains a two-input NAND gate followed by an inverter (Fig. 1.5). Standard cells are distributed in cell *libraries*, which are often provided at no cost by foundries and are pre-qualified for manufacturing.

AND			OR			INV		NAND			NOR		
IN1	IN2	OUT	IN1	IN2	OUT	IN	OUT	IN1	IN2	OUT	IN1	IN2	OUT
0	0	0	0	0	0	0	1	0	0	1	0	0	1
1	0	0	1	0	1	1	0	1	0	1	1	0	0
0	1	0	0	1	1			0	1	1	0	1	0
1	1	1	1	1	1			1	1	0	1	1	0

Fig. 1.5 Examples of common digital cells with their input and output behavior.

Standard cells are designed in multiples of a fixed cell height, with fixed locations of *power* (*VDD*) and *ground* (*GND*) ports. Cell widths vary depending on the transistor network implemented. As a consequence of this restricted layout style, all cells are placeable in rows such that power and ground supply nets are distributed by (horizontal) abutment (Fig. 1.6) The cells' signal ports may be located at the "upper" and "lower" cell boundaries or distributed throughout the cell area [1.3].

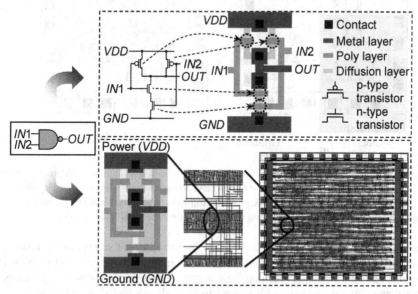

Fig. 1.6 Implementation of a NAND gate using CMOS technology (top), and as a standard cell (lower left), that can be embedded in a VLSI layout (lower right).

Since standard-cell placement has less freedom, the complexity of this design methodology is greatly reduced. This can decrease time-to-market at the cost of such metrics as power efficiency, layout density, or operating frequency, when compared to full-custom designs. Hence, standard-cell based designs, e.g., ASICs, address different market segments than full-custom designs, e.g., microprocessors, FPGAs, and memory products. A substantial initial effort is required to develop the cell library and qualify it for manufacturing.

The routing between standard-cell rows uses either *feedthrough* (empty) cells within rows or available routing tracks across rows (Fig. 1.7). When space between cell rows is available, such regions are called *channels*. These channels, along with the space above the cells, can also be used for routing. *Over-the-cell (OTC)* routing has become popular as a way to use multiple metal layers – up to 8-12 in modern designs – in current process technologies. This routing style is more flexible than traditional channel routing. If OTC routing is used, then adjacent standard-cell rows are typically not separated by routing channels but instead share either a power rail or a ground rail (Fig. 1.7 right). OTC routing is prevalent in industry today.

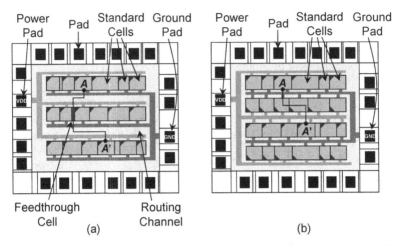

Fig. 1.7 (a) A standard-cell layout with net *A-A'* routed through a feedthrough cell and cell rows separated by channels; each row has its own power and ground rails. (b) A standard-cell layout with net *A-A'* routed using over-the-cell (OTC) routing; cell rows share power and ground rails, which requires alternating cell orientations. Designs with more than three metal layers use OTC routing.

Macro cells. Macro cells are typically larger pieces of logic that perform a reusable functionality. They can range from simple (a couple of standard cells) to highly complex (entire subcircuits reaching the scale of an embedded processor or memory block), and can vary greatly with respect to their shapes and sizes. In most cases, macro cells can be placed anywhere in the layout area with the goals of optimizing routing distance or electrical properties of the design.

Due to the increasing popularity of reusing optimized modules, macro cells, such as adders and multipliers, have become popular. In some cases, almost the entire functionality of a design can be assembled from pre-existing macros; this calls for *top-level assembly*, through which various subcircuits, e.g., analog blocks, standard-cell blocks, and "glue" logic, are combined with individual cells, e.g., buffers, to form the highest hierarchical level of a complex circuit (Fig. 1.8).

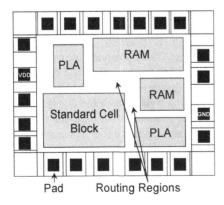

Fig. 1.8 Example layout with macro cells.

Gate arrays. Gate arrays are silicon chips with standard logic functionality, e.g., NAND and NOR, but no connections. The *interconnect* (routing) layers are added later after the chip-specific requirements are known. Since the gate arrays are not initially customized, they can be mass-produced. Then, the time-to-market of gate array-based designs is mostly constrained by the fabrication of interconnects. This makes gate array-based designs cheaper and faster to produce than standard cell-based or macro cell-based designs, particularly for low production volumes.

The layout of gate arrays is greatly restricted, so as to simplify modeling and design. Due to this limited freedom, wire-routing algorithms can be very straightforward. Only the following two tasks are needed.

- *Intracell* routing: creating a cell (logic block) by, e.g., connecting certain transistors to implement a NAND gate. Common gate connections are typically located in cell libraries.
- *Intercell* routing: connecting the logic blocks to form nets from the netlist.

During physical design of gate arrays, (1) cells are selected from what is available on the chip, and (2) since the demand for routing resources depends on the placement configuration, a bad placement may lead to failures at the routing stage. Several variants and extensions of the traditional gate-array style are now available.

Field-programmable gate arrays (FPGAs). In an FPGA, both logic elements and interconnects come prefabricated, but can be configured by users through switches (Fig. 1.9). *Logic elements* (*LEs*) are implemented by *lookup tables* (*LUTs*), each of which can represent any k-input Boolean function, e.g., $k = 4$ or $k = 5$. Interconnect is configured using *switchboxes* (*SBs*) that join wires in adjacent routing channels. Configurations of LUTs and switchboxes are read from external storage and stored locally in memory cells. The main advantage of FPGAs is their customization without the involvement of a fabrication facility. This dramatically reduces design costs, up-front investment and time-to-market. However, FPGAs typically run much slower and dissipate more power than ASICs [1.8]. Above certain production volumes, e.g., millions of chips, FPGAs become more expensive than ASICs since the non-recurring design and manufacturing costs of ASICs are amortized [1.2].

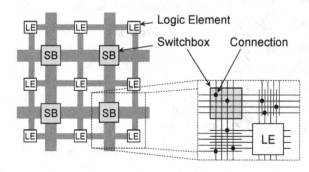

Fig. 1.9 Logic elements (LEs) are connected through switchboxes (SBs) that form a programmable routing network.

Structured-ASICs (channel-less gate arrays). A channel-less gate array s similar to an FPGA, except that the cells are usually not configurable. Unlike traditional gate arrays, sea of gate designs have many interconnect layers, removing the need for routing channels and thus improving density. The interconnects (sometimes, only via layers) are mask-programmed in the foundry, and are not field-programmable. The modern incarnation of the channel-less gate array is the *structured ASIC*.

1.4 Layout Layers and Design Rules

The gates and interconnects of an integrated circuit are formed using standard materials that are deposited and patterned on *layout layers*, with the layout patterns themselves conforming to *design rules* that ensure manufacturability, electrical performance and reliability.

Layout layers. Integrated circuits are made up of several different materials, the main ones being

– Single-crystal silicon substrate which is doped to enable construction of n- and p-channel transistors
– Silicon dioxide, which serves as an insulator
– Polycrystalline silicon or *polysilicon*, which forms transistor gates and can serve as an interconnect material
– Either aluminum or copper, which serves as metal interconnect

Silicon serves as the *diffusion layer*. The polysilicon and the aluminum and copper layers are collectively referred to as interconnect layers; the polysilicon is called *poly* and the remaining layers are called *Metal1*, *Metal2*, etc. (Fig. 1.10). *Vias* and *contacts* make connections between the different layers – vias connect metal layers and contacts connect poly and Metal1.

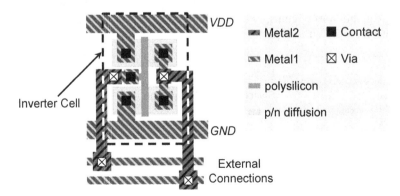

Fig. 1.10 The different layers for a simple inverter cell, showing external connections to the channel below and internal connections.

The wire resistance is usually given as *sheet resistance* in ohms per square ($\Omega/\square$). That is, for a given wire thickness, the resistance per square area remains the same – independent of the square size (a higher resistance for a longer length is compensated by the increased width of the square).[1] Hence, the resistance of any rectangular interconnect shape can be easily calculated as the number of unit-square areas multiplied by the sheet resistance of the corresponding layer.

Individual transistors are created by overlapping poly and diffusion layers. Cells, e.g., standard cells, are comprised of transistors but typically include one metal layer.

The routing between cells (Chaps. 5-7) is performed entirely within the metal layers. This is a non-trivial task – not only are poly and Metal1 mostly reserved for cells, but different layers have varying sheet resistances, which strongly affects timing characteristics. For a typical 0.35 µm CMOS process, the sheet resistance of poly is 10 $\Omega/\square$, that of the diffusion layer is approximately 3 $\Omega/\square$, and that of aluminum is 0.06 $\Omega/\square$. Thus, poly should be used sparingly, and most of the routing done in metal layers.

Routing through multiple metal layers requires vias. For the same 0.35 µm process, the typical resistance of a via between two metal layers is 6 Ω, while that of a contact is significantly higher – 20 Ω. As technology scales, modern copper interconnects become highly resistive due to smaller cross sections, grain effects that cause electron scattering, and the use of barrier materials to prevent reactive copper atoms from leaching into the rest of the circuit. In a typical 65 nm CMOS process, the sheet resistance of poly is 12 $\Omega/\square$, that of the diffusion layer is 17 $\Omega/\square$, and that of the copper Metal1 layer is 0.16 $\Omega/\square$. Via and contact resistances are respectively 1.5 Ω and 22 Ω in a typical 65 nm process.

Design rules. An integrated circuit is fabricated by shining laser light through *masks*, where each mask defines a certain layer pattern. For a mask to be effective, its layout pattern must meet specific technology constraints. These constraints, or *design rules*, ensure that (1) the design can be fabricated in the specified technology and (2) the design will be electrically reliable and feasible. Design rules exist both for each individual layer and for interactions across multiple layers. In particular, transistors require structural overlaps of poly and diffusion layers.

Though design rules are complex, they can be broadly grouped into three categories.

–   *Size rules*, such as *minimum width*: The dimensions of any component (shape), e.g., length of a boundary edge or area of the shape, cannot be smaller than given minimum values (*a* in Fig. 1.11). These values vary across different metal layers.

---

[1] Since [length/width] is dimensionless, sheet resistance is measured in the same units as resistance (ohms). However, to distinguish it from resistance, it is specified in ohms per square ($\Omega/\square$).

- *Separation rules*, such as *minimum separation*: Two shapes, either on the same layer (*b* in Fig. 1.11) or on adjacent layers (*c* in Fig. 1.11), must be a minimum (rectilinear or Euclidean diagonal) distance apart (*d* in Fig. 1.11).
- *Overlap rules*, such as *minimum overlap*: Two connected shapes on adjacent layers must have a certain amount of overlap (*e* in Fig. 1.11) due to inaccuracy of mask alignment to previously-made patterns on the wafer.

To enable technology scaling, fabrication engineers use a standard unit *lambda* (λ) to represent the minimum size of any design feature [1.9]. Thus, design rules are specified in multiples of λ, which facilitates grid-based layout with base length λ. Such a framework is very convenient, in that technology scaling only affects the value of λ. However, as the size of transistors decreases, the λ metric is becoming less meaningful, as some physical and electrical properties no longer follow such ideal scaling.

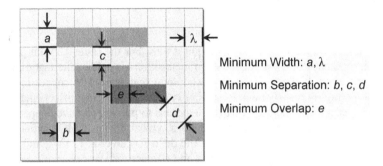

Minimum Width: *a*, λ

Minimum Separation: *b*, *c*, *d*

Minimum Overlap: *e*

Fig. 1.11 Several classes of design rules. The grid's granularity is λ, the smallest meaningful technology-dependent unit of length.

1.5 Physical Design Optimizations

Physical design is a complex *optimization problem* with several different objectives, such as minimum chip area, minimum wirelength, and minimum of vias. Common goals of optimization are to improve circuit performance, reliability, etc. How well the optimization goals are met determines the quality of the layout.

Different optimization goals (1) may be difficult to capture within algorithms and (2) may conflict with each other. However, tradeoffs among multiple goals can often be expressed concisely by an *objective function*. For instance, wire routing can optimize

$$w_1 \cdot A + w_2 \cdot L$$

where A is the chip area, L is the total wirelength, and w_1 and w_2 are *weights* that represent the relative importance of A and L. In other words, the weights influence the impact of each objective goal on the overall cost function. In practice, $0 \leq w_1 \leq 1$, $0 \leq w_2 \leq 1$, and $w_1 + w_2 = 1$.

During layout optimization, three types of *constraints* must be met.

— *Technology constraints* enable fabrication for a specific technology node and are derived from technology restrictions. Examples include minimum layout widths and spacing values between layout shapes.
— *Electrical constraints* ensure the desired electrical behavior of the design. Examples include meeting maximum timing constraints for signal delay and staying below maximum coupling capacitances.
— *Geometry (design methodology) constraints* are introduced to reduce the overall complexity of the design process. Examples include the use of preferred wiring directions during routing, and the placement of standard cells in rows.

As technology scales further, electrical effects have become increasingly significant. Thus, many types of electrical constraints have been introduced recently to ensure correct behavior. Various constraints not required at earlier technology nodes are necessary for modern designs. Such constraints may limit *coupling capacitance* to ensure *signal integrity*, prevent *electromigration* effects in interconnects, and prevent adverse temperature-related phenomena.

A basic challenge is that new electrical effects are not easily translated into new geometric rules for layout design. For instance, is signal delay best minimized by reducing total wirelength or by reducing coupling capacitance between the routes of different nets? Such a question is further complicated by the fact that routes on other metal layers, as well as their switching activity, also affect signal delay. Although only loose geometric rules can be defined, electrical properties can be accurately extracted from layout, and physical simulation enables precise estimation of timing, noise and power. This allows designers to assess the impact of layout optimizations.

In summary, difficulties encountered when optimizing layout include the following.

— Optimization goals may conflict with each other. For example, minimizing wirelength too aggressively can result in a congested region, and increase the number of vias.
— Constraints often lead to discontinuous, qualitative effects even when objective functions remain continuous. For example, the floorplan design might permit only some of the bits of a 64-bit bus to be routed with short wires, while the remaining bits must be detoured.
— Constraints, due to scaling and increased interconnect requirements, are tightening, with new constraint types added for each new technology node.

These difficulties motivate the following rules of thumb.

– Each design style requires its own custom flow. That is, there is no universal EDA tool that supports all design styles.
– When designing a chip, imposing geometric constraints can potentially make the problem easier at the expense of layout optimization. For instance, a row-based standard-cell design is much easier to implement than a full-custom layout, but the latter could achieve significantly better electrical characteristics.
– To further reduce complexity, the design process is divided into sequential steps. For example, placement and routing are performed separately, each with specific optimization goals and constraints that are evaluated independently.
– When performing fundamental optimizations, the choice is often between (1) an abstract model of circuit performance that admits a simple computation, or (2) a realistic model that is computationally intractable. When no efficient algorithm or closed-form expression is available to obtain a globally optimal solution, the use of heuristics is a valid and effective option (Sec. 1.6).

1.6 Algorithms and Complexity

A key criterion for assessing any algorithm is its *runtime complexity*, the time required by the algorithm to complete as a function of some natural measure of the problem size. For example, in block placement, a natural measure of problem size is the number of blocks to be placed, and the time $t(n)$ needed to place n blocks can be expressed as

$$t(n) = f(n) + c$$

where $f(0) = 0$ and c is a fixed amount of "overhead" that is required independently of input size, e.g., during initialization.

While other measures of algorithm complexity such as memory ("space") are also of interest, runtime is the most important complexity metric for IC physical design algorithms. Complexity is represented in an *asymptotic* sense, with respect to the input size n, using *big-Oh notation* or $O(...)$. Formally, the runtime $t(n)$ is *order* $f(n)$, written as $t(n) = O(f(n))$ when

$$\lim_{n \to \infty} \left| \frac{t(n)}{f(n)} \right| = k$$

where k is a real number. For example, if $t(n) = 7n! + n^2 + 100$, then $t(n) = O(n!)$ because $n!$ is the fastest growing term as $n \to \infty$. The real number k for $f(n) = n!$ is

$$k = \lim_{n \to \infty} \left| \frac{7n! + n^2 + 100}{n!} \right| = \lim_{n \to \infty} \left| \frac{7n!}{n!} + \frac{n^2}{n!} + \frac{100}{n!} \right| = 7 + 0 + 0 = 7$$

Placement problems and their associated computational complexities include

- Place n cells in a single row and return the wirelength: $O(n)$
- Given a single-row placement of n cells, determine whether the wirelength can be improved by swapping one pair of cells: $O(n^2)$
- Given a single-row placement of n cells, determine whether the wirelength can be improved by permuting a group of three cells at a time: $O(n^3)$
- Place n cells in a single row so as to minimize the wirelength: $O(n! \cdot n)$ with a naive algorithm

Example: Exhaustively Enumerating All Placement Possibilities
Given: n cells.
Task: find a linear (single-row) placement of n cells with minimum total wirelength by using exhaustive enumeration.

Solution:
The solution space consists of $n!$ placement options. If generating and evaluating the wire-length of each possible placement solution takes 1 microsecond (μs) and $n = 20$, the total time needed to find an optimal solution would be 77,147 years!

The first three placement tasks are considered *scalable*, since their complexities can be written as $O(n^p)$ or $O(n^p \log n)$, where p is a small integer, usually $p \in \{1,2,3\}$. Algorithms having complexities where $p > 3$ are often considered *not scalable*. Furthermore, the last problem is considerably more difficult and is impractical for even moderate values of n, despite the existence of clever algorithms. A number of important problems have best-known algorithm complexities that grow exponentially with n, e.g., $O(n!)$, $O(n^n)$, and $O(e^n)$. Many of these problems are known to be *NP-hard*,[2] and no polynomial-time algorithms are currently known that solve these problems. Thus, for such problems, no known algorithms can ensure, in a time-efficient manner, that they will return a *globally optimal* solution.

Chaps. 2-8 all deal with physical design problems that are NP-hard. For these problems, *heuristic algorithms* are used to find near-optimal solutions within practical runtime limits. In contrast to conventional algorithms, which are guaranteed to produce an optimal (valid) solution in a known amount of time, heuristics may produce inferior solutions. Algorithms that have poor worst-case complexity, but produce optimal solutions in all practical cases, are also considered heuristics. The primary goal of algorithm development for EDA is to construct heuristics that can quickly

---

[2]   NP stands for *non-deterministic polynomial time*, and refers to the ability to validate in polynomial time any solution that was "non-deterministically guessed". NP-hard problems are at least as hard as the most difficult NP problems. For further reading, see [1.4] and [1.7].

produce near-optimal solutions on large and complex commercial designs by exploiting the typical features of such designs. Such heuristics often incorporate conventional algorithms for subtasks that can be solved optimally.

Types of algorithms. Many physical design problems, e.g., placement and routing, are NP-hard, so solving them with optimal, worst-case polynomial-time algorithms is unlikely. Many heuristics have been developed for these problems, the quality of which can be assessed based on (1) runtime and (2) solution quality, measured by *suboptimality* (difference from optimal solutions).

Heuristic algorithms can be classified as

- *Deterministic*: All decisions made by the algorithm are repeatable, i.e., not random. One example of a deterministic heuristic is Dijkstra's shortest path algorithm (Sec. 5.6.3).
- *Stochastic*: Some decisions made by the algorithm are made randomly, e.g., using a pseudo-random number generator. Thus, two independent runs of the algorithm will produce two different solutions with high probability. One example of a stochastic algorithm is simulated annealing (Sec. 3.5.3).

In terms of structure, a heuristic algorithm can be

- *Constructive*: The heuristic starts with an initial, incomplete (partial) solution and adds components until a complete solution is obtained.
- *Iterative*: The heuristic starts with a complete solution and repeatedly improves the current solution until a preset termination criterion is reached.

Physical design algorithms often employ both constructive and iterative heuristics. For instance, a constructive heuristic can be used to first generate an initial solution, which is then refined by an iterative algorithm (Fig. 1.12).

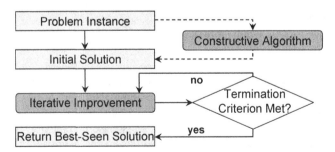

Fig. 1.12 Example of a heuristic that has both constructive and iterative stages. The constructive stage creates an initial solution that is refined by the iterative stage.

When the solution space is represented by a *graph* structure consisting of *nodes* and *edges* (Sec. 1.7), algorithms for search and optimization can be classified in another way, as described below. The graph structure can be *explicit*, as in wire routing, or *implicit*, with edges representing small differences between possible solutions, e.g., swapping a pair of adjacent standard cells in placement.

- *Breadth-first search* (*BFS*): When searching for goal node T from starting node S_0, the algorithm checks all adjacent nodes S_1. If the goal T is not found in S_1, the algorithm searches all of S_1's adjacent nodes S_2. This process continues, resembling expansion of a "wave-front", until T is found or all nodes have been searched.
- *Depth-first search* (*DFS*): From the starting node S_0, the algorithm checks nodes in order of increasing depth, i.e., traversing as far as possible and as soon as possible. In contrast to BFS, the next-searched node S_{i+1} is a neighbor of S_i unless all neighbors of S_i have already been searched, in which case the search backtracks to the highest-index location that has an unsearched neighbor. Thus, DFS traverses as far as possible as soon as possible.
- *Best-first search*: The direction of search is based on cost criteria, not simply on adjacency. Every step taken considers a current *cost* as well as the remaining cost to the goal. The algorithm always expands or grows from the current best known node or solution. An example is Dijkstra's algorithm (Sec. 5.6.3).

Finally, some algorithms used in physical design are *greedy*. An initial solution is transformed into another solution only if the new solution is strictly better than the previous solution. Such algorithms find *locally optimal* solutions. For further reading on the theory of algorithms and complexity, see [1.4].

Solution quality. Given that most physical design algorithms are heuristic in nature, the assessment of solution quality is difficult. If the optimal solution is known, then the heuristic solution can be judged by its *suboptimality* ε with respect to the optimal solution

$$\varepsilon = \frac{\left| cost(S_H) - cost(S_{opt}) \right|}{cost(S_{opt})}$$

where $cost(S_H)$ is the cost of the heuristic solution S_H and $cost(S_{opt})$ is the cost of the optimal solution S_{opt}. This notion applies to only a tiny fraction of design problems, in that optimal solutions are known only for small (or artificially-created) instances. On the other hand, *bounds* on suboptimality can sometimes be proven for particular heuristics, and can provide useful guidance.

When finding an optimal solution is impractical, as typical for modern designs, heuristic solutions are tested across a suite of *benchmarks*. These sets of (non-trivial) problem instances represent different corner cases, as well as common cases, and are inspired by either industry or academic research. They enable assessment of a given heuristic's scalability and solution quality against previously-obtained heuristic solutions.

1.7 Graph Theory Terminology

Graphs are heavily used in physical design algorithms to describe and represent layout topologies. Thus, a basic understanding of graph theory terminology is vital to understanding how the optimization algorithms work. The following is a list of basic terms; subsequent chapters will introduce specialized terminology.

A *graph* $G(V,E)$ is made up of two sets – the set of *nodes* or *vertices* (elements), denoted as V, and the set of *edges* (relations between the elements), denoted as E (Fig. 1.13(a)). The *degree* of a node is the number of its incident edges.

A *hypergraph* consists of nodes and *hyperedges*, with each hyperedge being a subset of two or more nodes. Note that a graph is a hypergraph in which every hyperedge has cardinality two. Hyperedges are commonly used to represent multi-pin nets or multi-point connections within circuit hypergraphs (Fig. 1.13(b)).

A *multigraph* (Fig. 1.13(c)) is a graph that can have more than one edge between two given nodes. Multigraphs can be used to represent varying *net weights*; an alternative is to use an edge-weighted graph representation, which is more compact and supports non-integer weights.

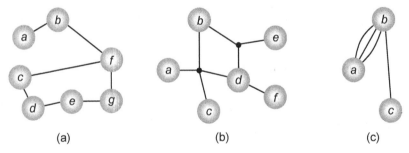

(a)   (b)   (c)

Fig. 1.13 (a) A graph with seven edges. (b) A hypergraph with three hyperedges having sizes four, three and two respectively. (c) A multigraph with four edges, where *a-b* has weight = 3.

A *path* between two nodes is an ordered sequence of edges from the start node to the end node (*a-b-f-g-e* in Fig. 1.13(a)).

A *cycle* (*loop*) is a closed path that starts and ends at the same node (*c-f-g-e-d-c* in Fig. 1.13(a)).

An *undirected graph* is a graph that represents only unordered node relations and does not have any directed edges. A *directed graph* is a graph where the direction of the edge denotes a specific ordered relation between two nodes. For example, a signal might be generated at the output pin of one gate and flow to an input pin of another gate – but not the other way around. Directed edges are drawn as arrows starting from one node and pointing to the other.

A directed graph is *cyclic* if it has at least one directed cycle (*c-f-g-d-c* in Fig. 1.14(a) or *a-b-a* in Fig. 1.14(b)). Otherwise, it is *acyclic*. Several important EDA algorithms operate on *directed acyclic graph* (*DAG*) representations of design data (Fig. 1.14(c)).

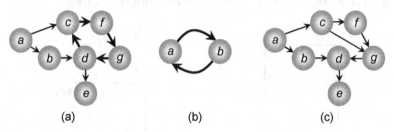

Fig. 1.14 (a) A directed graph with a cycle *c-f-g-d-c*. (b) A directed graph with a cycle *a-b-a*. (c) A directed acyclic graph (DAG).

A *complete graph* is a graph of *n* nodes with

$$\binom{n}{2} = \frac{n!}{2!(n-2)!} = \frac{n(n-1)}{2}$$

edges, one edge between each pair of nodes, i.e., each node is connected by an edge to every other node.

A *connected graph* is a graph with at least one path between each pair of nodes.

A *tree* is a graph whose *n* nodes are connected by *n* – 1 edges. There are two types of trees: undirected and directed (trees that have a *root*). Both types are shown in Fig. 1.15. In an undirected tree, any node with only one incident edge is a *leaf*. In a directed tree, the root has no incoming edge, and a leaf is a node with no outgoing edge. In a directed tree, there exists a unique path from the root to any other node.

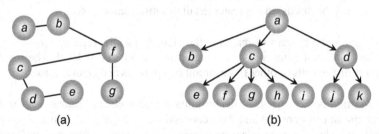

Fig. 1.15 (a) A undirected graph with leaves *a*, *e* and *g*, and maximum node degree 3. (b) A directed tree with root *a* and leaves *b* and *e-k*.

A *spanning tree* in a graph *G*(*V*,*E*) is a connected, acyclic subgraph *G'* contained within *G* that includes (spans) every node *v* ∈ *V*.

A *minimum spanning tree* (*MST*) is a spanning tree with the smallest possible sum of edge costs (i.e., edge lengths).

A *rectilinear minimum spanning tree* (*RMST*) is an MST where all edge lengths correspond to the *Manhattan* (rectilinear) distance metric (Fig. 1.16(a)).

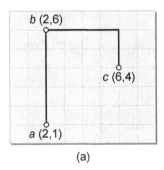 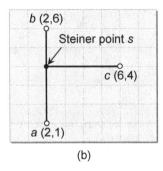

(a) (b)

Fig. 1.16 (a) Rectilinear minimum spanning tree (RMST) connecting points *a-c* with tree cost = 11. (b) Rectilinear minimum Steiner tree (RSMT) connecting points *a-c* with tree cost = 9.

The *Steiner tree*, named after *J. Steiner* (1796-1863), is a generalization of the spanning tree. In addition to the original nodes, Steiner trees have *Steiner points* (Fig. 1.16(b)). Edges are allowed to connect to these points as well as to the original nodes. The incorporation of Steiner points can reduce the total edge cost of the tree to below that of an RMST. A *Steiner minimum tree* (*SMT*) has minimum total edge cost over all Steiner trees. If constructed in the Manhattan plane using only horizontal and vertical segments, the SMT is a *rectilinear Steiner minimum tree* (*RSMT*).

1.8 Common EDA Terminology

The following is a brief list of important and common terms used in EDA. Many of these terms will be discussed in greater detail in subsequent chapters.

Logic design is the process of mapping the HDL (typically, register-transfer level) description to circuit gates and their connections at the netlist level. The result is usually a netlist of cells or other basic circuit components and connections.

Physical design is the process of determining the geometrical arrangement of cells (or other circuit components) and their connections within the IC layout. The cells' electrical and physical properties are obtained from library files and technology information. The connection topology is obtained from the netlist. The result of physical design is a geometrically and functionally correct representation in a standard file format such as GDSII Stream.

Physical verification checks the correctness of the layout design. This includes verifying that the layout

    –   Complies with all technology requirements – Design Rule Checking (DRC)
    –   Is consistent with the original netlist – Layout vs. Schematic (LVS)
    –   Has no antenna effects – Antenna Rule Checking
    –   Complies with all electrical requirements – Electrical Rule Checking (ERC)

A *component* is a basic functional element of a circuit. Examples include transistors, resistors, and capacitors.

A *module* is a circuit partition or a grouped collection of components.

A *block* is a module with a shape, i.e., a circuit partition with fixed dimensions.

A *cell* is a logical or functional unit built from various components. In digital circuits, cells commonly refer to gates, e.g., INV, AND-OR-INVERTER (AOI), NAND, NOR. In general, the term is used to refer to either standard cells or macros.

A *standard cell* is a cell with a pre-determined functionality. Its height is a multiple of a library-specific fixed dimension. In the standard-cell methodology, the logic design is implemented with standard cells that are arranged in rows.

A *macro cell* is a cell without pre-defined dimensions. This term may also refer to a large physical layout, possibly containing millions of transistors, e.g., an SRAM or CPU core, and possibly having discrete dimensions, that can be incorporated into the IC physical design.

A *pin* is an electrical terminal used to connect a given component to its external environment. At the level of block-to-block connections (internal to the IC), I/O pins are present on lower-level metal layers such as Metal1, Metal2 and Metal3. A *pad* is an electrical terminal used to connect externally to the IC. Often, bond pads are present on topmost metal layers and interface between external connections (such as to other chips) and internal connections.

A *layer* is a manufacturing process level in which design components are patterned. During physical design, circuit components are assigned to different layers, e.g., transistors are assigned to poly and active layers, while interconnects are assigned to poly and metal layers and are routed according to the netlist.

A *contact* is a direct connection between silicon (poly or another active level) and a metal layer, typically Metal1. Contacts are often used inside cells.

A *via* is a connection between metal layers, usually to connect routing structures on different layers.

A *net* or *signal* is a set of pins or terminals that must be connected to have the same potential.

Supply nets are power (*VDD*) and ground (*GND*) nets that provide current to cells.

A *netlist* is the collection of all signal nets and the components that they connect in a design, or, a list of all the nets and connecting pins of a subsection of the design. That is, netlists can be organized as (1) *pin-oriented* – each design component has a list of associated nets (Fig. 1.17 center), or (2) *net-oriented* – each net has a list of associated design components (Fig. 1.17 right). Netlists are created during logic synthesis and are a key input to physical design.

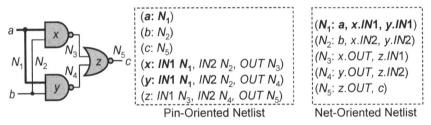

Pin-Oriented Netlist   Net-Oriented Netlist

Fig. 1.17 Pin-oriented (center) and net-oriented (right) netlist examples for the sample circuit (left).

A *net weight* w(*net*) is a numerical (typically integer) value given to a net *net* (or edge *edge*) to indicate its importance or criticality. Net weights are used primarily during placement, e.g., to minimize distance between cells that are connected by edges with high net weights, and routing, e.g., to set the priority of a net.

The *connectivity degree* or *connection cost* c(*i,j*) between cells *i* and *j* for un-weighted nets is the number of connections between *i* and *j*. With weighted nets, c(*i,j*) is the sum of the individual connection weight between *i* and *j*.

The *connectivity* c(*i*) of cell *cell$_i$* is given by

$$c(i) = \sum_{j=1}^{|V|} c(i, j)$$

where $|V|$ is the number of cells in the netlist, and c(*i,j*) is the connectivity degree between cells *i* and *j*. For example, cell *y* in Fig. 1.18 has c(*y*) = 5 if each net's weight equals 1.

A *connectivity graph* is a representation of the netlist as a graph. Cells, blocks and pads correspond to nodes, while their connections correspond to edges (Fig. 1.18). A *p*-pin net is represented by $\binom{p}{2}$ total connections between its nodes. Multiple edges between two nodes imply a stronger (*weighted*) connection.

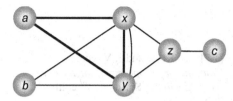

Fig. 1.18 Connectivity graph of the circuit in Fig. 1.17.

The *connectivity matrix* C is an $n \times n$ matrix that represents the circuit connectivity over n cells. Each element $C[i][j]$ represents the connection cost $c(i,j)$ between cells i and j (Fig. 1.19). Since C is symmetric, $C[i][j] = C[j][i]$ for $1 \le i, j \le m$. By definition, $C[i][i] = 0$ for $1 \le i \le m$, since a cell i does not meaningfully connect with itself.

	a	b	x	y	z	c
a	0	0	1	1	0	0
b	0	0	1	1	0	0
x	1	1	0	2	1	0
y	1	1	2	0	1	0
z	0	0	1	1	0	1
c	0	0	0	0	1	0

Fig. 1.19 The connectivity matrix of the circuit from Fig. 1.17. The entry $C[x][y] = C[y][x] = 2$ because both N_1 and N_2 contribute one unit.

The *Euclidean distance* metric between two points P_1 (x_1, y_1) and P_2 (x_2, y_2) corresponds to the length of the line segment between P_1 and P_2 (Fig. 1.20). In the coordinate plane, the Euclidean distance is

$$d_E(P_1, P_2) = \sqrt{(x_2 - x_1)^2 + (y_2 - y_1)^2}$$

The *Manhattan distance* metric between two points P_1 (x_1, y_1) and P_2 (x_2, y_2) is the sum of the horizontal and vertical displacements between P_1 and P_2 (Fig. 1.20). In the coordinate plane, the Manhattan distance is

$$d_M(P_1, P_2) = |x_2 - x_1| + |y_2 - y_1|$$

In Fig. 1.20, the Euclidean distance between P_1 and P_2 is $d_E(P_1, P_2) = 5$, whereas the Manhattan distance between P_1 and P_2 is $d_M(P_1, P_2) = 7$.

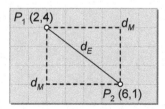

Fig. 1.20 Distance between two points P_1 and P_2 according to Euclidean (d_E) and Manhattan (d_M) distance metrics.

Chapter 1 References

[1.1]   C. J. Alpert, D. P. Mehta and S. S. Sapatnekar, eds., *Handbook of Algorithms for Physical Design Automation*, CRC Press, 2009.

[1.2]   D. Chen, J. Cong and P. Pan, "FPGA Design Automation: A Survey", *Foundations and Trends in EDA* 1(3) (2006), pp. 195-330.

[1.3]   D. Clein, *CMOS IC Layout: Concepts, Methodologies, and Tools*, Newnes, 1999.

[1.4]   T. Cormen, C. Leiserson, R. Rivest and C. Stein, *Introduction to Algorithms, 2ⁿᵈ Edition*, McGraw Hill, 2003.

[1.5]   *International Technology Roadmap for Semiconductors*, 2009 edition, www.itrs.net/.

[1.6]   H. Kaeslin, *Digital Integrated Circuit Design: From VLSI Architectures to CMOS Fabrication*, Cambridge University Press, 2008.

[1.7]   B. Korte and J. Vygen, *Combinatorial Optimization: Theory and Algorithms*, Springer, 3ʳᵈ edition, 2006.

[1.8]   I. Kuon and J. Rose, "Measuring the Gap Between FPGAs and ASICs", *IEEE Trans. on CAD* 26(2) (2007), pp. 203-215.

[1.9]   C. Mead and L. Conway, *Introduction to VLSI Systems*, Addison-Wesley, 1979.

[1.10] G. Moore, "Cramming More Components Onto Integrated Circuits", *Electronics* 38(8) (1965).

[1.11] L. Scheffer, L. Lavagno and G. Martin, eds., *EDA for IC Implementation, Circuit Design, and Process Technology*, CRC Press, 2006.

[1.12] L.-T. Wang, Y.-W. Chang and K.-T. Cheng, eds., *Electronic Design Automation: Synthesis, Verification, and Test (Systems on Silicon)*, Morgan Kaufmann, 2009.

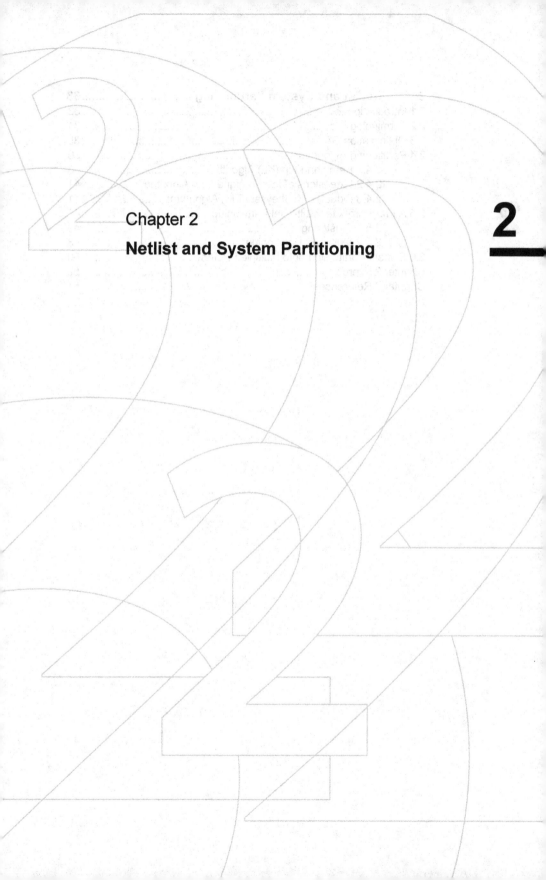

Chapter 2

Netlist and System Partitioning

2

2

2 Netlist and System Partitioning

The design complexity of modern integrated circuits has reached unprecedented scale, making full-chip layout, FPGA-based emulation and other important tasks increasingly difficult. A common strategy is to *partition* or divide the design into smaller portions, each of which can be processed with some degree of independence and parallelism. A *divide-and-conquer* strategy for chip design can be implemented by laying out each block individually and reassembling the results as *geometric partitions*. Historically, this strategy was used for manual partitioning, but became infeasible for large netlists. Instead, manual partitioning can be performed in the context of system-level modules by viewing them as single entities, in cases where hierarchical information is available. In contrast, automated *netlist partitioning* (Secs. 2.1-2.4) can handle large netlists and can redefine a physical hierarchy of an electronic system, ranging from boards to chips and from chips to blocks. Traditional netlist partitioning can be extended to *multilevel* partitioning (Sec. 2.5), which can be used to handle large-scale circuits and system partitioning on FPGAs (Sec. 2.6).

2.1 Introduction

A popular approach to decrease the design complexity of modern integrated circuits is to *partition* them into smaller *modules*. These modules can range from a small set of electrical components to fully functional integrated circuits (ICs). The partitioner divides the circuit into several subcircuits (partitions or blocks) while minimizing the number of connections between partitions, subject to design constraints such as maximum partition sizes and maximum path delay.

If each block is implemented independently, i.e., without considering other partitions, then connections between these partitions may negatively affect the overall design performance such as increased circuit delay or decreased reliability. Moreover, a large number of connections between partitions may introduce inter-block dependencies that hamper design productivity.[1] Therefore, the primary goal of partitioning is to divide the circuit such that the number of connections between subcircuits is minimized (Fig. 2.1). Each partition must also meet all design constraints. For example, the amount of logic in a partition can be limited by the size of an FPGA chip. The number of external connections of a partition may also be limited, e.g., by the number of I/O pins in the chip package.

---

[1] The empirical observation known as *Rent's rule* suggests a power-law relationship between the number of cells n_G and the number of external connections $n_P = t \cdot n_G{}^r$, for any subcircuit of a "well-designed" system. Here, t is the number of pins per cell and r, referred to as the *Rent's exponent* or the *Rent parameter*, is a constant < 1. In particular, Rent's rule quantifies the prevalence of short wires in ICs, which is consistent with a hierarchical organization.

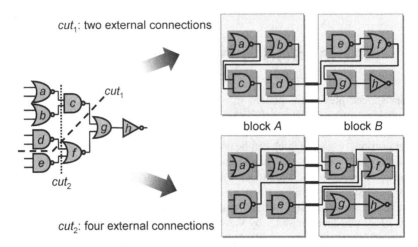

Fig. 2.1 Two different partitions induced by cuts cut_1 and cut_2, producing two and four external connections, respectively.

2.2 2.2 Terminology

The following are common terms relevant to netlist partitioning. Terminology relating to specific algorithms, such as the Kernighan-Lin algorithm (Sec. 2.4.1), will be introduced and defined in their respective sections.

A *cell* is any logical or functional unit built from components.

A *partition* or *block* is a grouped collection of components and cells.

The *k-way partitioning* problem seeks to divide a circuit into k partitions. Fig. 2.2 illustrates how the partitioning problem can be abstracted using a graph representation, where nodes represent cells, and edges represent connections between cells.

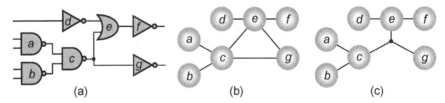

Fig. 2.2 (a) Sample circuit. (b) Possible graph representation. (c) Possible hypergraph representation.

Given a graph $G(V,E)$, for each node $v \in V$, $area(v)$ represents the area of the corresponding cell or module. For each edge $e \in E$, $w(e)$ represents the priority or weight, e.g., timing criticality, of the corresponding edge.

Though this chapter discusses the partitioning problem and partitioning algorithms within the graph context, logic circuits are more accurately represented using *hyper-graphs*, where each *hyperedge*[2] connects two or more cells. Many graph-based algorithms can be directly extended to hypergraphs.

The set of all partitions $|Part|$ is *disjoint* if each node $v \in V$ is assigned to *exactly one* of the partitions.

An edge between two nodes i and j is *cut* if i and j belong to different partitions A and B, i.e., $i \in A, j \in B$, and $(i,j) \in E$ (Fig. 2.3).

A *cut set* Ψ is the collection of all cut edges.

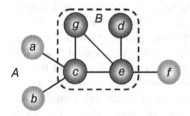

Fig. 2.3 A 2-way partitioning of the circuit in Fig. 2.2. A contains nodes a, b and f. B contains nodes c, d, e and g. Edges (a,c), (b,c) and (e,f) are cut. Edges (c,e), (c,g), (d,e) and (e,g) are not cut.

2.3 Optimization Goals

The most common partitioning objective is to minimize the number or total weight of cut edges while balancing the sizes of the partitions. If Ψ denotes the set of cut edges, the minimization objective is

$$\sum_{e \in \Psi} w(e)$$

Often, partition area is limited due to packing considerations and other boundary conditions implied by system hierarchy, chip size, or floorplan restrictions. For any subset of nodes $V' \subseteq V$, let $area(V')$ be the total area of all cells represented by the nodes of V'. *Bounded-size partitioning* enforces an upper bound UB on the total area of each partition V'. That is, $area(V_i) \leq UB_i$, where $V_i \subseteq V$, $i = 1, \ldots, k$, and k is the number of partitions. Often, a circuit must be divided evenly, with

$$area(V_i) = \sum_{v \in V_i} area(v) \leq \frac{1}{k} \sum_{v \in V} area(v) = \frac{1}{k} area(V)$$

---

[2] For convenience, hyperedges may be referred to as edges. However, graph edges are formally defined as node pairs.

For the special case where all nodes have unit area, the *balance criterion* is

$$|V_i| \le \left\lceil \frac{|V|}{k} \right\rceil$$

2.4 Partitioning Algorithms

Circuit partitioning, like many other combinatorial optimization problems discussed in this book, is *NP-hard*. That is, as the problem size grows linearly, the effort needed to find an optimal solution grows faster than any polynomial function. To date, there is no known polynomial-time, globally optimal algorithm for balance-constrained partitioning (Sec. 1.6). However, several efficient heuristics were developed in the 1970s and 1980s. These algorithms find high-quality circuit partitioning solutions and in practice are implemented to run in low-order polynomial time – the Kernighan-Lin (KL) algorithm (Sec. 2.4.1), its extensions (Sec. 2.4.2) and the Fiduccia-Mattheyses (FM) algorithm (Sec. 2.4.3). Additionally, optimization by *simulated annealing* can be used to solve particularly difficult partitioning formulations. In general, stochastic hill-climbing algorithms require more than polynomial time to produce high-quality solutions, but can be accelerated by sacrificing solution quality. In practice, simulated annealing is rarely competitive.

▶ **2.4.1 Kernighan-Lin (KL) Algorithm**

The Kernighan-Lin (KL) algorithm performs partitioning through iterative-improvement steps. It was proposed by *B. W. Kernighan* and *S. Lin* in 1970 [2.6] for *bipartitioning* ($k = 2$) graphs, where every node has unit weight. This algorithm has been extended to support k-way partitioning ($k > 2$) as well as cells with arbitrary areas (Sec. 2.4.2).

Introduction. The KL algorithm operates on a graph representation of the circuit, where nodes (edges) represent cells (connections between cells). Formally, let a graph $G(V,E)$ have $|V| = 2n$ nodes, where each node $v \in V$ has the same weight, and each edge $e \in E$ has a non-negative edge weight. The KL algorithm partitions V into two disjoint subsets A and B with minimum cut cost and $|A| = |B| = n$.

The KL algorithm is based on exchanging (swapping) pairs of nodes, each node from a different partition. The two nodes that generate the highest reduction in cut size are swapped. To prevent immediate move reversal (*undo*) and subsequent infinite loops, the KL algorithm *fixes* nodes after swapping them. Fixed nodes cannot be swapped until they are released, i.e., become *free*.

Execution of the KL algorithm proceeds in *passes*. Typically, the first pass or iteration begins with an arbitrary initial partition. In a given pass, after all nodes become fixed, the algorithm determines the prefix of the sequence of swaps within this pass that produces the largest *gain*, i.e., reduction of cut cost. All nodes included in this sequence are moved accordingly. The pass finishes by releasing all fixed nodes, so that all nodes are once again free. In each subsequent pass, the algorithm starts with the two partitions from the previous pass. All possible pair swaps are then reevaluated. If no improvement is found during a given pass, the algorithm terminates.

Terminology. The following terms are specifically relevant to the KL algorithm.

The *cut size* or *cut cost* of a graph with either unweighted or uniform-weight edges is the number of edges that have nodes in more than one partition. With weighted edges, the cut cost is the sum of the weights of all cut edges.

The cost $D(v)$ of moving a node $v \in V$ in a graph from partition A to B is

$$D(v) = |E_B(v)| - |E_A(v)|$$

where $E_B(v)$ is the set of v's incident edges that are cut by the cut line, and $E_A(v)$ is the set of v's incident edges that are not cut by the cut line. High costs ($D > 0$) indicate that the node should move, while low costs ($D < 0$) indicate that the node should stay within the same partition.

The *gain* $\Delta g(a,b)$ of swapping a pair of nodes a and b is the improvement in overall cut cost that would result from the node swap. A positive gain ($\Delta g > 0$) means that the cut cost is decreased, while a negative gain ($\Delta g < 0$) means that the cut cost is increased. The gain of swapping two nodes a and b is

$$\Delta g(a,b) = D(a) + D(b) - 2c(a,b)$$

where $D(a)$ and $D(b)$ are the respective costs of nodes a and b, and $c(a,b)$ is the connection weight between a and b. If an edge exists between a and b, then $c(a,b) = $ the edge weight between a and b. Otherwise, $c(a,b) = 0$.

Notice that simply adding $D(a)$ and $D(b)$ when calculating Δg assumes that an edge is cut (uncut) *before* the swap and will be uncut (cut) *after* the swap. However, this does not apply if the nodes are connected by an edge e, as it will be cut both before and after the swap. Therefore, the term $2c(a,b)$ corrects for this overestimation of gain from the swap.

The *maximum positive gain* G_m corresponds to the best prefix of m swaps within the swap sequence of a given pass. These m swaps lead to the partition with the minimum cut cost encountered during the pass. G_m is computed as the sum of Δg values over the first m swaps of the pass, with m chosen such that G_m is maximized.

$$G_m = \sum_{i=1}^{m} \Delta g_i$$

Within a pass of the KL algorithm, the moves are only used to find the move sequence $<1 \ldots m>$ and G_m. The moves are then applied after these have been found.

Algorithm. The Kernighan-Lin algorithm progresses as follows.

Kernighan-Lin Algorithm
Input: graph $G(V,E)$ with $|V| = 2n$
Output: partitioned graph $G(V,E)$

1.	(A,B) = INITIAL_PARTITION(G)	// arbitrary initial partition
2.	$G_m = \infty$	
3.	**while** $(G_m > 0)$	
4.	$i = 1$	
5.	$order = \varnothing$	
6.	**foreach** (node $v \in V$)	
7.	$status[v] = FREE$	// set every node as free
8.	$D[v] = COST(v)$	// compute $D(v)$ for each node
9.	**while** (!IS_FIXED(V))	// while all cells are not fixed, select free
10.	$(\Delta g_i,(a_i,b_i))$ = MAX_GAIN(A,B)	// node a_i from A and free node b_i from
		// B so that $\Delta g_i = D(a_i) + D(b_i) - 2c(a_i,b_i)$
		// is maximized
11.	ADD$(order,(\Delta g_i,(a_i,b_i)))$	// keep track of swapped cells
12.	TRY_SWAP(a_i,b_i,A,B)	// move a_i to B, and move b_i to A
13.	$status[a_i] = FIXED$	// mark a_i as fixed
14.	$status[b_i] = FIXED$	// mark b_i as fixed
15.	**foreach** (free node v_f connected to a_i and b_i)	
16.	$D[v_f] = COST(v_f)$	// compute and update $D(v_f)$
17.	$i = i + 1$	
18.	(G_m,m) = BEST_MOVES$(order)$	// swap sequence $1 \ldots m$ that
		// maximizes G_m
19.	**if** $(G_m > 0)$	
20.	CONFIRM_MOVES$(order,m)$	// execute move sequence

First, partition the input graph G into two arbitrary partitions A and B (line 1), and set the maximum positive gain value G_m to ∞ (line 2). During each pass (line 3), for each node $v \in V$, compute the cost $D(v)$ and set v to free (lines 4-6). Then, while all nodes are not fixed (line 9), select, swap, and fix two free nodes a_i and b_i from A and B, respectively, such that Δg_i is maximized (lines 10-14). Update $D(v)$ for all free nodes that are adjacent to a_i and b_i (lines 15-16).

After all nodes have been fixed, find the move sequence $<1 \ldots m>$ that maximizes G_m (line 18). If G_m is positive, execute the move sequence (lines 19-20), and perform another pass (lines 3-20). Otherwise, terminate the algorithm.

Example: KL Algorithm
Given: initial partition of nodes *a-h* (right).
Task: perform the first pass of the KL algorithm.

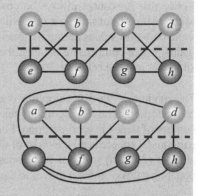

Solution:
Initial cut cost = 9.
Compute $D(v)$ costs for all free nodes *a-h*.
$D(a) = 1$, $D(b) = 1$, $D(c) = 2$, $D(d) = 1$,
$D(e) = 1$, $D(f) = 2$, $D(g) = 1$, $D(h) = 1$
$\Delta g_1 = D(c) + D(e) - 2c(c,e) = 2 + 1 - 0 = 3$
Swap and fix nodes *c* and *e*.

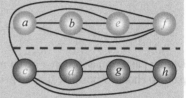

Update $D(v)$ costs for all free nodes connected to
newly swapped nodes *c* and *e*: *a, b, d, f, g* and *h*.
$D(a) = -1$, $D(b) = -1$, $D(d) = 3$,
$D(f) = 2$, $D(g) = -1$, $D(h) = -1$
$\Delta g_2 = D(d) + D(f) - 2c(d,f) = 3 + 2 - 0 = 5$
Swap and fix nodes *d* and *f*.

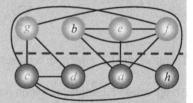

Update $D(v)$ costs for all free nodes connected to
newly swapped nodes *d* and *f*: *a, b, g* and *h*.
$D(a) = -3$, $D(b) = -3$, $D(g) = -3$, $D(h) = -3$
$\Delta g_3 = D(a) + D(g) - 2c(a,g) = -3 + -3 - 0 = -6$
Swap and fix nodes *a* and *g*.

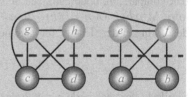

Update $D(v)$ costs for all free nodes connected to
newly swapped nodes *a* and *g*: *b* and *h*.
$D(b) = -1$, $D(h) = -1$
$\Delta g_4 = D(b) + D(h) - 2c(b,h) = -1 + -1 - 0 = -2$
Swap and fix nodes *b* and *h*.

Compute maximum positive gain G_m.
$\quad\quad G_1 = \Delta g_1 = 3$
$\quad\quad G_2 = \Delta g_1 + \Delta g_2 = 3 + 5 = 8$
$\quad\quad G_3 = \Delta g_1 + \Delta g_2 + \Delta g_3 = 3 + 5 + -6 = 2$
$\quad\quad G_4 = \Delta g_1 + \Delta g_2 + \Delta g_3 + \Delta g_4 = 3 + 5 + -6 + -2 = 0$

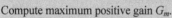

$G_m = 8$ with $m = 2$. Since $G_m > 0$, the first $m = 2$ swaps are executed: (c,e) and (d,f). Additional passes are performed until $G_m \leq 0$.

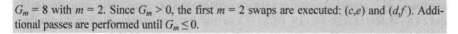

The KL algorithm does not always find an optimal solution, but performs reasonably well in practice on graphs with up to several hundred nodes. In general, the number of passes grows with the size of the graph, but in practice improvement often ceases after four passes.

Notice that Δg cannot always be positive: after all nodes have been swapped between two partitions, the cut cost will be exactly the same as the initial cut cost, so some best-gain values during the pass can be negative. However, since other moves (gains) might compensate for this, the entire pass should be completed, computing all moves until all cells are fixed.

The runtime of the KL algorithm is dominated by two factors – gain updates and pair selection. The KL algorithm selects n pairs of nodes to swap, where n is the number of nodes in each partition. For each node v, the required time to update the gains and compare is on the order of $O(n)$. That is, after swapping a_i and b_i in move i, at most $(2n - 2i)$ gains of free nodes must be updated. Therefore, the time spent updating gains over the n moves in a pass is at most

$$\sum_{i=1}^{n} 2n - 2i = O(n^2)$$

During pair comparison in a given move i, there are as many as $(n - i + 1)^2 = O(n^2)$ pairs to choose from. The time to perform n pair comparisons is bounded by

$$\sum_{i=1}^{n} (n - i + 1)^2 = O(n^3)$$

Therefore, the KL algorithm requires a total of $O(n^2) + O(n^3) = O(n^3)$ time.

An optimized KL implementation has $O(n^2 \log n)$ runtime complexity. To speed up pair comparison, node pairs can be sorted ahead of time. Since the goal is to maximize $\Delta g(a,b) = D(a) + D(b) - 2c(a,b)$, the gains of each node move can be sorted in descending order. That is, for each node $a \in A$, order the gains $D(a)$ such that

$$D(a_1) \geq D(a_2) \geq \ldots \geq D(a_{n-i+1})$$

Similarly, for each node $b \in B$, order the gains $D(b)$ such that

$$D(b_1) \geq D(b_2) \geq \ldots \geq D(b_{n-i+1})$$

Then, evaluate pairwise gains, starting with the first elements from both lists. A clever order of evaluation – exploiting advanced data structures and bounded node degrees [2.3] – allows the pair evaluation process to stop once a pair of gains $D(a_j)$ and $D(b_k)$ is found with $D(a_j) + D(b_k)$ is less than the best previously-found gain (no better pair-swap can exist). In practice, the best pair-swap at the kth move can be found in $O(n - k)$ time after sorting the free node gains in $O((n - k) \log (n - k))$ time [2.2]. The time required to perform pair comparison is thus reduced from $O(n^2)$ time to $O(n \log n)$ time.

▶ **2.4.2 Extensions of the Kernighan-Lin Algorithm**

To accommodate *unequal partition sizes* $|A| \neq |B|$, arbitrarily split the nodes among the two partitions A and B, where one partition contains $\min(|A|,|B|)$ nodes and the other $\max(|A|,|B|)$ nodes. Apply the KL algorithm with the restriction that only $\min(|A|,|B|)$ node pairs can be swapped.

To accommodate *unequal cell sizes* or *unequal node weights*, assign a *unit area* that denotes the smallest cell area, i.e., the greatest common divisor of all cell areas. All unequal node sizes are then cast as integer multiples of the unit area. Each node portion (all parts of an original node that was split up) is connected to each of its counterparts by *infinite-weight*, i.e., high-priority edges. Apply the KL algorithm.

To perform k-way partitioning, arbitrarily assign all $k \cdot n$ nodes to partitions such that each partition has n nodes. Apply the KL 2-way partitioning algorithm to all possible pairs of subsets (1 and 2, 2 and 3, etc.) until none of the consecutive KL applications obtains any improvement on the cut size.

▶ **2.4.3 Fiduccia-Mattheyses (FM) Algorithm**

Given a graph $G(V,E)$ with nodes and *weighted* edges, the goal of (bi)partitioning is to assign all nodes to disjoint partitions, so as to minimize the total cost (weight) of all cut nets while satisfying partition size constraints. The Fiduccia-Mattheyses (FM) algorithm is a partitioning heuristic, published in 1982 by *C. M. Fiduccia* and *R. M. Mattheyses* [2.4], offers substantial improvements over the KL algorithm.

–   *Single* cells are moved independently instead of swapping pairs of cells. Thus, this algorithm is more naturally applicable to partitions of unequal size or the presence of initially fixed cells.
–   Cut costs are extended to include *hypergraphs* (Sec. 1.8). Thus, all nets with two or more pins can be considered. While the KL algorithm aims to minimize cut costs based on *edges*, the FM algorithm minimizes cut costs based on *nets*.
–   The area of each individual cell is taken into account.
–   The selection of cells to move is much faster. The FM algorithm has runtime complexity of $O(|Pins|)$ per pass, where $|Pins|$ is the total number of pins, defined as the sum of all edge degrees $|e|$ over all edges $e \in E$.

Introduction. The FM algorithm is typically applied to large circuit netlists. For this section, all nodes and subgraphs are referred to as *cells* and *blocks*, respectively.

The FM move selection process is similar to that of the KL algorithm, with the underlying objective being to minimize cut cost. However, the FM algorithm computes the gain of each individual cell move, rather than that of each pair-swap. Like the KL algorithm, the FM algorithm selects the best prefix of moves from within a *pass*.

During an FM pass, once a cell is moved, it becomes fixed and cannot be moved for the remainder of the pass. The cells that are moved during the FM algorithm are denoted by the sequence $<c_1 \ldots c_m>$, whereas the KL algorithm swaps the first m pairs.

Terminology. The following definitions are relevant to the FM algorithm.

A net is *cut* if its cells occupy more than one partition. Otherwise, the net is *uncut*.

The *cut set* of a partition *part* is the set of all nets that are marked as *cut* within *part*.

The *gain* $\Delta g(c)$ for cell c is the change in the cut set size if c moves. The higher the gain $\Delta g(c)$, the higher is the priority to move the cell c to the other partition. Formally, the cell gain is defined as

$$\Delta g(c) = FS(c) - TE(c)$$

where $FS(c)$ is the number of nets connected to c but not connected to any other cells within c's partition, i.e., *cut nets that connect only to c*, and $TE(c)$ is the number of *uncut* nets connected to c. Informally, $FS(c)$ is like a *moving* force – the higher the $FS(c)$ value, the stronger the pull to move c to the other partition. $TE(c)$ is like a *retention force* – the higher the $TE(c)$ value, the stronger the desire to remain in the current partition.

The *maximum positive gain* G_m of a pass is the cumulative cell gain of m moves that produce a minimum cut cost. G_m is determined by the maximum sum of cell gains Δg over a prefix of m moves in a pass

$$G_m = \sum_{i=1}^{m} \Delta g_i$$

As in the KL algorithm, all moves in a pass are used to determine G_m and the move sequence $<c_1 \ldots c_m>$. Only at the end of the pass, i.e., after determining G_m and the corresponding m moves, are the cell positions updated (moved).

The *ratio factor* is the relative balance between the two partitions with respect to cell area. This ratio factor is used to prevent all cells from clustering into one partition. The ratio factor r is defined as

$$r = \frac{area(A)}{area(A) + area(B)}$$

where $area(A)$ and $area(B)$ are the total respective areas of partitions A and B, and

$$area(A) + area(B) = area(V)$$

where $area(V)$ is the total area of all cells $c \in V$, and is defined as

$$area(V) = \sum_{c \in V} area(c)$$

The *balance criterion* enforces the ratio factor. To ensure feasibility, the maximum cell area $area_{max}(V)$ must be taken into account. A partitioning of V into two partitions A and B is said to be *balanced* if

$$r \cdot area(V) - area_{max}(V) \le area(A) \le r \cdot area(V) + area_{max}(V)$$

A *base cell* is a cell c that has maximum cell gain $\Delta g(c)$ among all free cells, and whose move does not violate the balance criterion.

The *pin distribution* of a net *net* is given as a pair $(A(net), B(net))$, where $A(net)$ is the number of pins in partition A and $B(net)$ is the number of pins in partition B.

A net *net* is *critical* if it contains a cell c whose move changes the cut state of *net*. A critical net is either contained completely within a partition, or has exactly one of its cells in one partition and all of its other cells in the other partition. If *net* is critical, then either $A(net) = 0$, $A(net) = 1$, $B(net) = 0$, or $B(net) = 1$ must hold (Fig. 2.4).

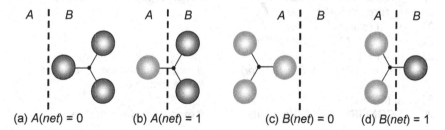

(a) $A(net) = 0$ (b) $A(net) = 1$ (c) $B(net) = 0$ (d) $B(net) = 1$

Fig. 2.4 Cases when a net *net* is critical. (a) $A(net) = 0$. (b) $A(net) = 1$. (c) $B(net) = 0$. (d) $B(net) = 1$.

Critical nets simplify the calculation of cell gains. Only cells belonging to critical nets need to be considered in the gain computation, as it is only for such nets that the movement of *a single cell* can change the cut state. B. *Krishnamurthy* [2.8] generalized the concept of criticality and improved the FM algorithm such that it comprehends how many cell moves nets are away from being critical. This results in a *gain vector* for each cell instead of a single *gain value* – the ith element of the gain vector for a free cell c_f records how many nets will become i cell moves away from being uncut if c_f moves.

The *from-block F* and *to-block T* define the direction in which a cell moves. That is, a cell moves from F to T.

Algorithm. The Fiduccia-Mattheyses algorithm progresses as follows.

Fiduccia-Mattheyses Algorithm
Input: graph $G(V,E)$, ratio factor r
Output: partitioned graph $G(V,E)$

1.	(lb,ub) = BALANCE_CRITERION(G,r)	// compute balance criterion
2.	(A,B) = PARTITION(G)	// initial partition
3.	$G_m = \infty$	
4.	**while** $(G_m > 0)$	
5.	$i = 1$	
6.	$order = \varnothing$	
7.	**foreach** (cell $c \in V$)	// for each cell, compute the
8.	$\Delta g[i][c]$ = FS(c) − TE(c)	// gain for current iteration,
9.	$status[c]$ = $FREE$	// and set each cell as free
10.	**while** (!IS_FIXED(V))	// while there are free cells, find
11.	$cell$ = MAX_GAIN$(\Delta g[i],lb,ub)$	// the cell with maximum gain
12.	ADD$(order,(cell,\Delta g[i]))$	// keep track of cells moved
13.	$critical\_nets$ = CRITICAL_NETS$(cell)$	// critical nets connected to $cell$
14.	**if** $(cell \in A)$	// if $cell$ belongs to partition A,
15.	TRY_MOVE$(cell,A,B)$	// move $cell$ from A to B
16.	**else**	// otherwise, if $cell$ belongs to B,
17.	TRY_MOVE$(cell,B,A)$	// move $cell$ from B to A
18.	$status[cell]$ = $FIXED$	// mark $cell$ as fixed
19.	**foreach** (net $net \in critical\_nets$)	// update gains for critical cells
20.	**foreach** (cell $c \in net, c \neq cell$)	
21.	**if** $(status[c]$ == $FREE)$	
22.	UPDATE_GAIN$(\Delta g[i][c])$	
23.	$i = i + 1$	
24.	(G_m,m) = BEST_MOVES$(order)$	// move sequence $c_1 \ldots c_m$ that
		// maximizes G_m
25.	**if** $(G_m > 0)$	
26.	CONFIRM_MOVES$(order,m)$	// execute move sequence

First, compute the balance criterion (line 1) and partition the graph (line 2). During each pass (line 4), determine the gain for each free cells c based on the number of incident cut nets $FS(c)$ and incident uncut nets $TE(c)$ (lines 7-8). After all the cell gains have been determined, select the base cell maximizes gain subject to satisfying the balance criterion (line 11). If multiple cells meet these two criteria then break ties, e.g., to keep partition sizes as close as possible to the middle of the ranges given by the balance criterion.

After the base cell has been selected, move it from its current partition to the other partition, and mark it as fixed (line 14-18). Of the remaining free cells, only those connected to the base cell by critical nets need to have their gains updated (lines 19-22). The gain of an unmoved cell can only change if it is connected to a moved cell (a base cell). Therefore, gain values change only when a net goes from occupying two partitions to one (positive gain) or vice-versa (negative gain). After updating the necessary cell gains, select the next base cell that has maximum gain value and satisfies the balance criterion. Continue this process until there are no remaining free cells (lines 10-23). Once all cells have been fixed, find the best prefix of the move sequence $<c_1 \ldots c_m>$ (line 24) to achieve maximum positive gain

$$G_m = \sum_{i=1}^{m} \Delta g_i$$

As long as $G_m > 0$, execute the move sequence, free all cells, and begin a new pass. Otherwise, terminate the algorithm (lines 25-26).

Example: FM Algorithm
Given: (1) weighted cells a-e, (2) ratio factor $r = 0.375$, (3) nets N_1-N_5, and (4) initial partition (right).

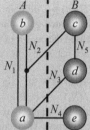

$area(a) = 2$ $area(b) = 4$ $area(c) = 1$ $area(d) = 4$ $area(e) = 5$
$N_1 = (a,b)$ $N_2 = (a,b,c)$ $N_3 = (a,d)$ $N_4 = (a,e)$ $N_5 = (c,d)$
Task: perform the first pass of the FM algorithm.

Solution:
Compute the balance criterion.
$r \cdot area(V) - area_{max} \leq area(A) \leq r \cdot area(V) + area_{max}(V)$
$r \cdot area(V) - area_{max}(V) = 0.375 \cdot 16 - 5 = 1$, $r \cdot area(V) + area_{max}(V) = 0.375 \cdot 16 + 5 = 11$.
Size range of A: $1 \leq area(A) \leq 11$.

Compute the gains of each cell a-e.
Nets N_3 and N_4 are cut: $FS(a) = 2$. Net N_1 is connected to a but is not cut: $TE(a) = 1$.
$\Delta g_1(a) = 2 - 1 = 1$. The cut size will be reduced if a moves from A to B.
Similarly, the gains of the remaining cells are computed.

b:	$FS(b) = 0$	$TE(b) = 1$	$\Delta g_1(b) = -1$
c:	$FS(c) = 1$	$TE(c) = 1$	$\Delta g_1(c) = 0$
d:	$FS(d) = 1$	$TE(d) = 1$	$\Delta g_1(d) = 0$
e:	$FS(e) = 1$	$TE(e) = 0$	$\Delta g_1(e) = 1$

Select the base cell. Possible base cells are a and e.
Balance criterion after moving a: $area(A) = area(b) = 4$.
Balance criterion after moving e: $area(A) = area(a) + area(b) + area(e) = 11$.

Both moves respect the balance criterion, but a is selected, moved, and fixed as a result of the tie-breaking criterion described above.

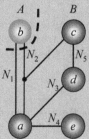

Update gains for all free cells that are connected by critical nets to a. To determine whether a given net is critical, the number of cells associated with that net in each partition is counted before and after the move.

For a given net net, let the number of cells in the from-block and to-block $before$ the move be $F(net)$ and $T(net)$, respectively. Let the number of cells in the from-block and to-block $after$ the move be $F'(net)$ and $T'(net)$, respectively. If any of these values is 0 or 1, then net is critical. For nets N_1, N_2, N_3, N_4, $T(N_1) = 0$ and $T(N_2) = T(N_3) = T(N_4) = 1$. Therefore, cells b, c, d and e need updating. The gain values do not have to be computed explicitly but can be derived from $T(net)$.

If $T(net) = 0$, all gain values of free cells connected to net increase by 1. Since $T(N_1) = 0$, cell b has updated gain value of $\Delta g_1(b) = \Delta g_1(b) + 1$. That is, net N_1 (connected to cell b) has increased the cut set of the partition. The increase in $\Delta g(b)$ reflects the motivation to move cell b. Since net N_1 is now cut, moving cell b is justified.

If $T(net) = 1$, all gain values of free cells connected to net decrease by 1. Since $T(N_2) = T(N_3) = T(N_4) = 1$, cells c, d and e have updated gain values of $\Delta g_1(c,d,e) = \Delta g_1(c,d,e) - 1$. That is, nets N_2, N_3 and N_4 (connected to cells c, d and e, respectively) have decreased the cut set of the partition. This reduction in $\Delta g_1(c,d,e)$ reflects the motivation to not move cells c, d and e. Similarly, when $F'(net) = 0$, all cell gains connected to net are reduced by 1, and when $F'(net) = 1$, all cell gains connected to n are increased by 1.

The updated Δg values are

b:	$FS(b) = 2$	$TE(b) = 0$	$\Delta g_1(b) = 2$
c:	$FS(c) = 0$	$TE(c) = 1$	$\Delta g_1(c) = -1$
d:	$FS(d) = 0$	$TE(d) = 2$	$\Delta g_1(d) = -2$
e:	$FS(e) = 0$	$TE(e) = 1$	$\Delta g_1(e) = -1$

Iteration $i = 1$
Partitions: $A_1 = \{b\}$, $B_1 = \{a,c,d,e\}$, with fixed cells $\{a\}$.

Iteration $i = 2$
Cell b has maximum gain $\Delta g_2 = 2$, $area(A) = 0$, balance criterion is violated.
Cell c has next maximum gain $\Delta g_2 = -1$, $area(A) = 5$, balance criterion is met.
Cell e has next maximum gain $\Delta g_2 = -1$, $area(A) = 9$, balance criterion is met.
Move cell c, updated partitions: $A_2 = \{b,c\}$, $B_2 = \{a,d,e\}$, with fixed cells $\{a,c\}$.

Iteration $i = 3$
Gain values: $\Delta g_3(b) = 1$, $\Delta g_3(d) = 0$, $\Delta g_3(e) = -1$.
Cell b has maximum gain $\Delta g_3 = 1$, $area(A) = 1$, balance criterion is met.
Move cell b, updated partitions: $A_3 = \{c\}$, $B_3 = \{a,b,d,e\}$, with fixed cells $\{a,b,c\}$.

Iteration $i = 4$
Gain values: $\Delta g_4(d) = 0$, $\Delta g_4(e) = -1$.
Cell d has maximum gain $\Delta g_4 = 0$, $area(A) = 5$, balance criterion is met.
Move cell d, updated partitions: $A_4 = \{c,d\}$, $B_4 = \{a,b,e\}$, with fixed cells $\{a,b,c,d\}$.

Iteration $i = 5$
Gain values: $\Delta g_5(e) = -1$.
Cell e has maximum gain $\Delta g_5 = -1$, $area(A) = 10$, balance criterion is met.
Move cell e, updated partitions: $A_5 = \{c,d,e\}$, $B_5 = \{a,b\}$, all cells fixed.

Find best move sequence $<c_1 \ldots c_m>$

$$G_1 = \Delta g_1 = 1$$
$$G_2 = \Delta g_1 + \Delta g_2 = 0$$
$$G_3 = \Delta g_1 + \Delta g_2 + \Delta g_3 = 1$$
$$G_4 = \Delta g_1 + \Delta g_2 + \Delta g_3 + \Delta g_4 = 1$$
$$G_5 = \Delta g_1 + \Delta g_2 + \Delta g_3 + \Delta g_4 + \Delta g_5 = 0$$

Maximum positive cumulative gain $G_m = \sum_{i=1}^{m} \Delta g_i = 1$ found in iterations 1, 3 and 4.

The move prefix $m = 4$ is selected due to the better balance ratio ($area(A) = 5$); the four cells a, b, c and d are then moved.

Result of Pass 1: Current partitions: $A = \{c,d\}$, $B = \{a,b,e\}$, cut cost reduced from 3 to 2.

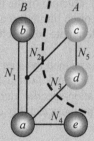

Pass 2 is left as a further exercise (Exercise 3). In this pass, the cut cost is reduced from 2 to 1. In Pass 3, no further improvement can be made. Thus, the partition found in Pass 2 is the final solution returned by the FM algorithm with a cut cost of 1.

2.5 A Framework for Multilevel Partitioning   2.5

Among the partitioning techniques discussed so far, the Fiduccia-Mattheyses heuristic offers the best tradeoff between solution quality and runtime. In particular, it is much faster than other techniques and, in practice, finds better partitions given the same amount of time. Unfortunately, if the partitioned hypergraph includes more than several hundred nodes, the FM algorithm may terminate with a high net cut or make a large number of passes, each producing minimal improvement.

To improve the scalability of netlist partitioning, the FM algorithm is typically embedded into a *multilevel* framework that consists of several distinct steps. First, during the *coarsening* phase, the original "flat" netlist is hierarchically clustered. Second, FM is applied to the clustered netlist. Third, the netlist is partially unclustered during the *uncoarsening* phase. Fourth, during the *refinement* phase, FM is applied incrementally to the partially unclustered netlist. The third and fourth steps are continued until the netlist is fully unclustered. In other words, FM is applied to the partially unclustered netlist and the solution is unclustered further – with this process repeating until the solution is completely flat.

For circuits with hundreds of thousands of gates, the multilevel framework dramatically improves runtime because many of the FM calls operate on smaller netlists, and each incremental FM call has a relatively high-quality initial solution. Furthermore, solution quality is improved, as applying FM to clustered netlists allows the algorithm to reassign entire clusters to different partitions where appropriate.

▶ 2.5.1 Clustering

To construct a coarsened netlist, groups of tightly-connected nodes can be *clustered*, absorbing connections between these nodes (Fig. 2.5). The remaining connections between clusters retain the overall structure of the original netlist. In specific applications, the size of each cluster is often limited so as to prevent *degenerate* clustering, where a single large cluster dominates other clusters.

When merging nodes, a cluster is assigned the sum of the weights of its constituent nodes. As closed-form objective functions for clustering are difficult to formulate, clustering is performed by application-specific algorithms. Additionally, clustering must be performed quickly to ensure the scalability of multilevel partitioning.

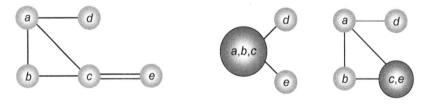

Fig. 2.5 An initial graph (left), and possible clusterings of the graph (right).

▶ 2.5.2 Multilevel Partitioning

Multilevel partitioning techniques begin with a *coarsening* phase in which the input graph G is clustered into a smaller graph G', which, in turn, can also be clustered into another graph G'', and so on. Let l be the number of levels, i.e., times, that G goes through the coarsening stage. Each node at level l represents a cluster of nodes at level $l + 1$. For large-scale applications, the *clustering ratio*, i.e., the average number of nodes per cluster, is often 1.3 (*Hypergraph Partitioning and Clustering*, Chap. 61 in [2.5]). For a graph with $|V|$ nodes, the number of levels can be estimated as

$$\lceil \log(|V| / v_0) \rceil$$

where v_0 is the number of nodes in the most-clustered graph (level 0), and the base of the logarithm is the clustering ratio. Clustering with this ratio is repeated until the graph is small enough to be processed efficiently by the FM partitioning algorithm. In practice, v_0 is typically 75-200, an instance size for which the FM algorithm is nearly optimal. For netlists with more than 200 nodes, multilevel partitioning techniques improve both runtime and solution quality. Academic multilevel partitioners include *hMetis* [2.6], available as a binary, and *MLPart* [2.1], available as open-source software.

The most-clustered netlist is partitioned using the FM algorithm. Then, its cluster-to-partition assignment is projected onto the larger, partially-unclustered netlist one level below. This is done by assigning every subcluster to the partition to which its parent cluster was assigned.

Using that partition as a starting configuration, subclusters in the partially-unclustered graph can be moved by the FM algorithm from one partition to another to improve cost while satisfying balance criteria (*refinement*). The process continues until the bottom-level netlist is refined (Fig. 2.6).

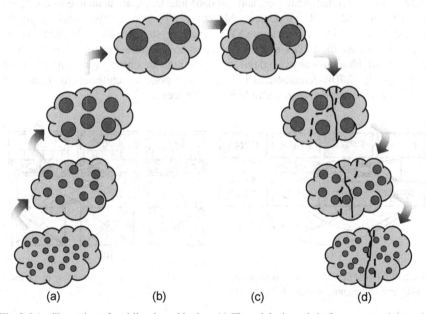

(a)   (b)   (c)   (d)

Fig. 2.6 An illustration of multilevel partitioning. (a) The original graph is first coarsened through several levels. (b) The graph after coarsening. (c) After coarsening, a heuristic partition is found of the most-coarsened graph. (d) That partition is then projected onto the next-coarsest graph (dotted line) and then refined (solid line). Projection and refinement continue until a partitioning solution for the original graph is found.

2.6 System Partitioning onto Multiple FPGAs

System implementation using field-programmable gate arrays (FPGAs), such as those manufactured by *Xilinx* or *Altera*, is an increasingly important application of partitioning. There are two main reasons for this trend. First, FPGA-based system prototyping allows products to meet shorter time-to-market windows, since embedded software development and design debugging can proceed concurrently with hardware design, rather than having to wait until the packaged dies arrive from the foundry. Second, with increased non-recurring engineering costs (mask sets and probe cards) in advanced technology nodes, products with lower production volumes become economically feasible only when implemented using field-programmable devices. However, field-programmability (e.g., using SRAM-based lookup tables to implement reconfigurable logic and interconnect) comes at the cost of density, speed and power. Hence, even if a system easily fits onto a single ASIC, its prototype may require multiple FPGA devices.

Functionally, FPGA-based systems may be viewed as logic (implemented using reprogrammable FPGAs) and interconnects (implemented using field-programmable interconnect chips, or FPICs). Many system components, including embedded processor cores, embedded memories, and standard interfaces, are available as configurable IPs on modern FPGA devices. Moreover, FPICs themselves can be implemented using FPGAs. An example FPGA-based system topology is illustrated in Fig. 2.7(a), where the FPGA and FPIC devices are connected using a *Clos* network topology, which allows any two devices to communicate directly (or a small number of hops). Fig. 2.7(b) demonstrates how a typical system architecture of logic and memory can be mapped onto multiple FPGA devices.

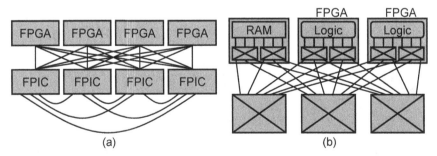

Fig. 2.7 (a) Reconfigurable system with multiple FPGA and FPIC devices. (b) Mapping of a typical system architecture onto multiple FPGAs.

Key challenges for multi-way system partitioning onto FPGAs include (1) low utilization of FPGA gate capacity because of hard I/O pin limits, (2) low clock speeds due to long interconnect delays between multiple FPGAs, and (3) long runtimes for the system partitioning process itself. This section discusses the associated algorithmic challenges in physical design that are unique to system implementation with multiple FPGAs.

Variant multi-way partitioning formulations. Multi-way partitioning for system prototyping seeks to minimize the number of FPGA devices needed while taking into account both area constraints, i.e., the partitions must each fit into individual FPGAs, and I/O constraints, i.e., each FPGA has a fixed number of pins. In contrast to the single-chip context, a small change in balance or cut size can make a feasible solution infeasible. Thus, a challenge for partitioning algorithms is to achieve high utilization of the FPGA devices while meeting all I/O constraints.

Once the number of FPGA devices has been determined, the secondary optimization objective is to minimize the amount of communication between the connected devices. Adopting general techniques for minimizing the net cut to FPGA-based architectures can significantly improve the overall speed of the system. However, the traditional net cut objective does not distinguish whether the gates of a five-pin net are split across two, three, four of five FPGA devices. However, splitting a net across k FPGA devices consumes k I/O pins. Hence, k should be minimized first.

Variant placement formulations. The reprogrammable nature of FPGAs allows systems to be implemented as true reconfigurable computing machines, where device configuration bits are updated to match the implemented logic to the required computation. This induces an extra "dimension" to the problems of logic partitioning and placement – the solution must explicitly evolve through time, i.e., through the course of the computation.

System implementation degrees of freedom. More performance optimizations are available, and needed, at the system level than during place-and-route. System prototyping may need to explore netlist transformations such as *cloning* (Sec. 8.5.3) and *retiming* (Sec. 8.6) in order to minimize cut size (I/O usage) or system cycle time. Such transformations are needed as inter-device delays can be relatively large and because devices are often I/O-limited. *L.-T. Liu et al.* [2.9] proposed a partitioning algorithm that permits logic replication to minimize both cut size and clock cycle of sequential circuits. Given a netlist $G = (V,E)$, their approach chooses two modules as seeds s and t and constructs a "replication graph" that is twice the size of the original circuit. This graph has the special property that a type of directed minimum cut yields the *replication cut*, i.e., a decomposition of V into S, T and R, where $s \in S$, $t \in T$ and $R = V - S - T$ is the replicated logic, that is optimal. A directed version of the Fiduccia-Mattheyses algorithm (Sec. 2.4.3) is used to find a heuristic directed minimum cut in the replication graph.

"Flow-based" multi-way partitioning method. To decompose a system into multiple devices, *C.-W. Yeh et al.* [2.10] proposed a "flow-based" algorithm inspired by the relationship between *multi-commodity flow* [2.2] and the traditional problem of min-cut partitioning. The algorithm constructs a flow network wherein each signal net initially corresponds to an edge with unit flow cost. To visualize this, one can imagine a network of roads, where each road corresponds to a signal net in the netlist, and where driving along each individual road requires a unit toll. Two random modules in the network are chosen, and the shortest (lowest-cost) path between them is computed. A constant $\gamma < 1$ is added to the flow for each net in the shortest path, and the cost for every net in the path is incremented. Adjusting the cost penalizes paths through congested areas and forces alternative shortest paths. This random shortest path computation is repeated until every path between the chosen pair of modules passes through at least one "saturated" net (in the analogy, this would be a "congested road"). The set of saturated nets induces a multi-way partitioning in which two modules belong to the same cluster if and only if there is a path of unsaturated nets between them. A second phase of the algorithm makes the multi-way partitioning more balanced. Since this approach has efficient runtime and is easily parallelizable, it is well-suited for large-scale multi-way system partitioning.

Commercial tools for partitioning large systems onto FPGAs typically base their algorithms on the multilevel extensions of the FM algorithm (Sec. 2.4.3). While it is not always possible to modify such algorithms to track relevant partitioning objectives directly, these algorithms often produce reasonable initial partitions when guided by the net cut objective and its variants.

Chapter 2 Exercises

Exercise 1: KL Algorithm

The graph to the right (nodes *a-f*) can be optimally partitioned using the Kernighan-Lin algorithm. Perform the first pass of the algorithm. The dotted line represents the initial partitioning. Assume all nodes have the same weight and all edges have the same priority.

Note: Clearly describe each step of the algorithm. Also, show the resulting partitioning (after one pass) in graphical form.

Exercise 2: Critical Nets and Gain During the FM Algorithm

(a) For cells *a-i*, determine the critical nets connected to these cells and which critical nets remain after partitioning. For the first iteration of the FM algorithm, determine which cells would need to have their gains updated due to a move. Hint: It may be helpful to prepare a table with one row per move that records (1) the cell moved, (2) critical nets before the move, (3) critical nets after the move, and (4) which cells require a gain update.

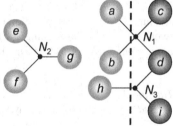

(b) Determine $\Delta g(c)$ for each cell $c \in V$.

Exercise 3: FM Algorithm

Perform Pass 2 of the FM algorithm example given in Sec. 2.4.3. Clearly describe each step. Show the result of each iteration in both numerical and graphical form.

Exercise 4: System and Netlist Partitioning

Explain key differences between partitioning formulations used for FPGA-based system emulation and traditional min-cut partitioning.

Exercise 5: Multilevel FM Partitioning

List and explain the advantages that a multilevel framework offers compared to the FM algorithm alone.

Exercise 6: Clustering

Consider a partitioned netlist. Clustering algorithms covered in this chapter do not take a given partitioning into account. Explain how these algorithms can be modified such that each new cluster is consistent with one of the initial partitions.

Chapter 2 References

[2.1]   A. E. Caldwell, A. B. Kahng and I. L. Markov, "Design and Implementation of Move-Based Heuristics for VLSI Hypergraph Partitioning", *J. Experimental Algorithmics* 5 (2000), pp. 1-21.

[2.2]   T. Cormen, C. Leiserson, R. Rivest and C. Stein, *Introduction to Algorithms, 2nd Edition*, McGraw Hill, 2003.

[2.3]   S. Dutt, "New Faster Kernighan-Lin-Type Graph-Partitioning Algorithms", *Proc. Intl. Conf. on CAD*, 1993, pp. 370-377.

[2.4]   C. M. Fiduccia and R. M. Mattheyses, "A Linear-Time Heuristic for Improving Network Partitions", *Proc. Design Autom. Conf.*, 1982, pp. 175-181.

[2.5]   T. F. Gonzalez, ed., *Handbook of Approximation Algorithms and Metaheuristics*, CRC Press, 2007.

[2.6]   G. Karypis, R. Aggarwal, V. Kumar and S. Shekhar, "Multilevel Hypergraph Partitioning: Application in VLSI Domain", *Proc. Design Autom. Conf.*, 1997, pp. 526-529.

[2.7]   B. W. Kernighan and S. Lin, "An Efficient Heuristic Procedure for Partitioning Graphs", *Bell Sys. Tech. J.* 49(2) (1970), pp. 291-307.

[2.8]   B. Krishnamurthy, "An Improved Min-Cut Algorithm for Partitioning VLSI Networks", *IEEE Trans. on Computers* 33(5) (1984), pp. 438-446.

[2.9]   L.-T. Liu, M.-T. Kuo, C.-K. Cheng and T. C. Hu, "A Replication Cut for Two-Way Partitioning", *IEEE Trans. on CAD* 14(5) (1995), pp. 623-630.

[2.10] C.-W. Yeh, C.-K. Cheng and T.-T. Y. Lin, "A General Purpose, Multiple-Way Partitioning Algorithm", *IEEE Trans. on CAD* 13(12) (1994), pp. 1480-1488.

Chapter 3
Chip Planning

3

3

3 Chip Planning

Chip planning deals with large modules such as caches, embedded memories, and intellectual property (IP) cores that have known areas, fixed or changeable shapes, and possibly fixed locations. When modules are not clearly specified, chip planning relies on *netlist partitioning* (Chap. 2) to identify such modules in large designs. Assigning shapes and locations to circuit modules during chip planning produces *blocks,* and enables early estimates of interconnect length, circuit delay and chip performance. Such early analysis can identify modules that need improvement. Chip planning consists of three major stages (1) *floorplanning,* (2) *pin assignment,* and (3) *power planning.*

Recall from Chap. 2 that a gate-level or RTL netlist can be automatically partitioned into *modules.* Alternatively, such modules can be extracted from a hierarchical design representation. Large chip modules are laid out as *blocks* or rectangular shapes (Fig. 3.1). *Floorplanning* determines the locations and dimensions of these shapes, based on the areas and aspect ratios of the modules so as to optimize chip size, reduce interconnect and improve timing. *Pin assignment* connects outgoing signal nets to block pins. *I/O placement* finds the locations for the chip's input and output pads, often at the periphery of the chip. This step is (ideally) performed before floorplanning, but locations can be updated during and after floorplanning. *Power planning* builds the *power supply network,* i.e., *power* and *ground* nets, so as to ensure that each block is provided with appropriate supply voltage. The results of partitioning and chip planning greatly affect subsequent design steps.

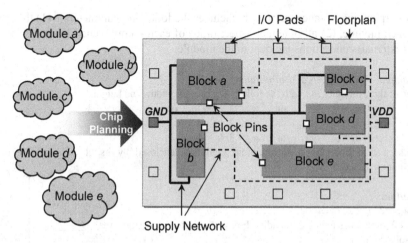

Fig. 3.1 Chip planning and relevant terminology. A module is a cluster of logic with a known area. Once it has been assigned shape or dimensions, it becomes a block. Connections between blocks are implemented through internal pins but are not shown in the figure.

3.1 Introduction to Floorplanning

Before the floorplanning stage, the design is split into individual circuit modules. A module becomes a *rectangular block* after it is assigned dimensions or a shape.[1] These blocks can be either *hard* or *soft*. The dimensions and areas of hard blocks are fixed. For a soft block, the area is fixed but the aspect ratio can be changed, either continuously or in discrete steps. The entire arrangement of blocks, including their positions, is called a *floorplan*. In large designs, individual modules may also be floorplanned in a recursive top-down fashion, but it is common to focus on one hierarchical level of floorplanning at a time. In this case, the floorplan of the highest level is called the *top-level floorplan*.

The floorplanning stage ensures that (1) every chip module is assigned a shape and a location, so as to facilitate gate placement, and (2) every pin that has an external connection is assigned a location, so that internal and external nets can be routed.

The floorplanning stage determines the external characteristics – fixed dimensions and external pin locations – of each module. These characteristics are necessary for the subsequent placement (Chap. 4) and routing (Chaps. 5-7) steps, which determine the internal characteristics of the blocks. Floorplan optimization involves multiple degrees of freedom; while it includes some aspects of placement (finding locations) and connection routing (pin assignment), module shape optimization is unique to floorplanning. Floorplanning with *hard blocks* is particularly relevant when reusing pre-existing blocks, including *intellectual property* (*IP*). Mathematically, this problem can be viewed as a constrained case of floorplanning with *soft* parameters, but in practice, it may require specialized computational techniques.

A floorplanning instance commonly includes the following parameters – (1) the area of each module, (2) all potential aspect ratios of each module, and (3) the netlist of all (external) connections incident to the module.

Example: Floorplan Area Minimization
Given: three modules *a-c* with the following potential widths and heights.
a: $w_a = 1, h_a = 4$ or $w_a = 4, h_a = 1$ or $w_a = 2, h_a = 2$
b: $w_b = 1, h_b = 2$ or $w_b = 2, h_b = 1$
c: $w_c = 1, h_c = 3$ or $w_c = 3, h_c = 1$
Task: find a floorplan with minimum total area enclosed by its global bounding box (definition in Sec. 3.2).

Solution:
a: $w_a = 2, h_a = 2$ *b:* $w_b = 2, h_b = 1$ *c:* $w_c = 1, h_c = 3$
This floorplan has a global bounding box with minimum possible area (9 square units).
A possible arrangement of the blocks (right).

---

[1] Blocks can also be rectilinear.

3.2 Optimization Goals in Floorplanning

Floorplan design optimizes both the *locations* and the *aspect ratios* of the individual blocks, using simple objective functions to capture practically desirable floorplan attributes. This section introduces several objective functions for floorplanning. Goals for pin assignment are discussed in Sec. 3.6, and goals for power planning are described in Sec. 3.7.

Area and shape of the global bounding box. The *global bounding box* of a floorplan is the minimum axis-aligned (isothetic) rectangle that contains all floorplan blocks. The area of the global bounding box represents the area of the top-level floorplan (the full design) and directly impacts circuit performance, yield, and manufacturing cost. Minimizing the area of the global bounding box involves finding (x,y) locations, as well as shapes, of the individual modules such that they pack densely together.

Beyond area minimization, another optimization objective is to keep the aspect ratio of the global bounding box as close as possible to a given target value. For instance, due to manufacturing and package size considerations, a square chip (aspect ratio $\approx$ 1) may be preferable to a non-square chip. To this end, the shape flexibility of the individual modules can be exploited. Area and aspect ratio of the global bounding box are interrelated, and these two objectives are often considered together.

Total wirelength. Long connections between floorplan blocks may increase signal propagation delays in the design. Therefore, layout of high-performance circuits seeks to shorten such interconnects. Switching the logic value carried by a particular net requires energy dissipation that grows with wire capacitance. Therefore, power minimization may also seek to shorten all routes. A third context for wirelength minimization involves routability and manufacturing cost. When the total length of all connections is too high or when the connections are overly dense in a particular region, there may not be enough routing resources to complete all connections. Although circuit blocks may be spread further apart to add new routing tracks, this increases chip size and manufacturing cost, and may further increase net length.

To simplify computation of the total wirelength of the floorplan, one option is to connect all nets to the centers of the blocks. Although this technique does not yield a precise wirelength estimate, it is relatively accurate for medium-sized and small blocks, and enables rapid interconnect evaluation [3.17]. Two common approaches to model connectivity within a floorplan are to use (1) a connection matrix C (Sec. 1.8) representing the union of all nets, along with pairwise distances between blocks, or (2) a minimum spanning tree for each net (Sec. 5.6). Using the first model, the total connection length $L(F)$ of the floorplan F is estimated as

$$L(F) = \sum_{i,j \in F} C[i][j] \cdot d_M(i,j)$$

where element $C[i][j]$ of C is the degree of connectivity between blocks i and j, and $d_M(i,j)$ is the Manhattan distance between the center points of i and j (Sec. 1.8).

Using the second model, the total connection length $L(F)$ is estimated as

$$L(F) = \sum_{net \in F} L_{MST}(net)$$

where $L_{MST}(net)$ is the minimal spanning tree cost of net *net*.

In practice, more sophisticated wirelength objectives are often used. The center-pin location assumption may be improved by using actual pin locations [3.17]. The Manhattan distance wiring cost approximation may be improved by using pin-to-pin shortest paths in a graph representation of available routing resources. This can reflect not only distance, but routing congestion, signal delay, obstacles, and routing channels as well. With these refinements, wiring estimation in a floorplan relies on the construction of heuristic Steiner minimum trees in a weighted graph (Chap. 5).

Combination of area and total wirelength. To reduce both the total area *area(F)* and the total wirelength $L(F)$ of floorplan F, it is common to minimize

$$\alpha \cdot area(F) + (1 - \alpha) \cdot L(F)$$

where the parameter $0 \le \alpha \le 1$ gives the relative importance between *area(F)* and $L(F)$. Other terms, such as the aspect ratio of the floorplan, can be added to this objective function [3.3]. In practice, the area of the global bounding box may be a constraint rather than an optimization objective. This is appropriate when the package size and its cavity dimensions are fixed, or when the global bounding box is part of a higher-level system organization across multiple chips. In this case, wirelength and other objectives are optimized subject to the constraint that the floorplan fits inside a prescribed global bounding box (the *fixed-outline floorplanning problem*).

Signal delays. Until the 1990s, transistors that made up logic gates were the greatest contributor to chip delay. Since then, due to different delay scaling rates, interconnect delays have gradually become more important, and increasingly determine the chip's achievable clock frequency. Delays of long wires are particularly sensitive to the locations and shapes of floorplan blocks. A desirable quality of a floorplan is short wires connecting its blocks, such that all timing requirements are met. Often, critical paths and nets are given priority during floorplanning so that they span short distances.

Floorplan optimization techniques have been developed that use *static timing analysis* (Sec. 8.2.1) to identify the interconnects that lie on *critical paths*. If timing is violated, i.e., path delays exceed given constraints, the floorplan is modified to shorten critical interconnects and meet timing constraints [3.7].

3.3 Terminology

A *rectangular dissection* is a division of the chip area into a set of *blocks* or non-overlapping rectangles.

A *slicing floorplan* is a rectangular dissection obtained by repeatedly dividing each rectangle, starting with the entire chip area, into two smaller rectangles using a horizontal or vertical cut line.

A *slicing tree* or *slicing floorplan tree* is a binary tree with k leaves and $k-1$ internal nodes, where each leaf represents a block and each internal node represents a horizontal or vertical cut line (Fig. 3.2). This book uses a standard notation, denoting horizontal and vertical cuts by H and V, respectively. A key characteristic of the slicing tree is that each internal node has exactly two children. Therefore, every slicing floorplan can be represented by at least one slicing tree.

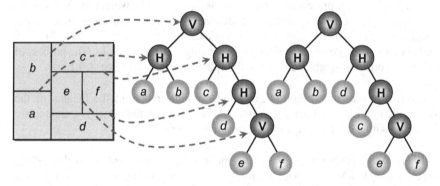

Fig. 3.2 A slicing floorplan of blocks *a-f* and two possible corresponding slicing trees.

A *non-slicing floorplan* is a floorplan that *cannot* be formed by a sequence of only vertical or horizontal cuts in a parent block. The smallest example of a non-slicing floorplan without wasted space is the *wheel* (Fig. 3.3).

 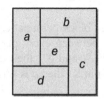

Fig. 3.3 Two different minimal non-slicing floorplans, also known as *wheels*.

A *floorplan tree* is a tree that represents a hierarchical floorplan (Fig. 3.4). Each leaf node represents a block while each internal node represents either a *horizontal cut* (*H*), a *vertical cut* (*V*), or a *wheel* (*W*). The *order* of the floorplan tree is the number of its internal (non-leaf) nodes.

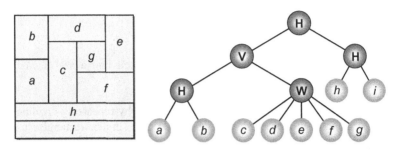

Fig. 3.4 A hierarchical floorplan (left) and its corresponding floorplan tree of order five (right). The internal node *W* represents a wheel with blocks *c-g*.

A *constraint-graph pair* is a floorplan representation that consists of two directed graphs – *vertical constraint graph* and *horizontal constraint graph* – which capture the relations between block positions. A constraint graph consists of edges connecting $n + 2$ weighted nodes – one source node s, one sink node t, and n block nodes $v_1, v_2, \ldots, v_n$ representing n blocks $m_1, m_2, \ldots, m_n$. The weight $w(v_i)$ of a block node v_i, $1 \leq i \leq n$, represents the size (relevant dimension) of the corresponding block; the weights of the source node $w(s)$ and sink node $w(t)$ are zero. Fig. 3.5 illustrates a sample floorplan and its corresponding constraint graphs. The process of deriving a constraint-graph pair from a floorplan is discussed in Sec. 3.4.1.

In a *vertical constraint graph* (*VCG*), node weights represent the heights of the corresponding blocks. Two nodes v_i and v_j, with corresponding blocks m_i and m_j, are connected with a directed edge from v_i to v_j if m_i is below m_j.

In a *horizontal constraint graph* (*HCG*), node weights represent the widths of the corresponding blocks. Two nodes v_i and v_j, with corresponding blocks m_i and m_j, are connected with a directed edge from v_i to v_j if m_i is to the left of m_j.

The longest path in the VCG corresponds to the minimum vertical extent required to pack the blocks (floorplan height). Similarly, the longest path in the HCG corresponds to the minimum horizontal extent required (floorplan width). In Fig. 3.5, a longest path is shown in both the HCG and VCG.

A *sequence pair* is an ordered pair (S_+, S_-) of block permutations. Together, the two permutations represent geometric relations between every pair of blocks a and b. Specifically, if a appears before b in both S_+ and S_-, then a is to the left of b. Otherwise, if a appears before b in S_+ but not in S_-, then a is above b.

$S_+: <...a...b...>$ $S_-: <...a...b...>$ if block a is *left of* block b

$S_+: <...a...b...>$ $S_-: <...b...a...>$ if block a is *above* block b

Fig. 3.5 illustrates a sample floorplan and its corresponding sequence pair. The process to derive a sequence pair from a floorplan is discussed in Sec. 3.4.2.

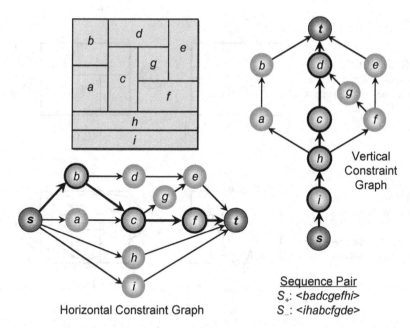

Fig. 3.5 A floorplan of blocks *a-i* (top left), its vertical constraint graph (top right), its horizontal constraint graph (bottom left), and its sequence pair (bottom right).

3.4 Floorplan Representations   3.4

This section discusses how to convert (1) a floorplan into a constraint-graph pair, (2) a floorplan into a sequence pair, and (3) a sequence pair into a floorplan. Note that modern sequence pair algorithms do not compute constraint graphs explicitly. However, the constraint graphs offer a useful, intermediate representation between floorplans and sequence pairs for a better conceptual understanding.

▶ 3.4.1 Floorplan to a Constraint-Graph Pair

A floorplan can be converted to a constraint-graph pair in three steps. First, for each constraint graph, create a block node v_i for each of the n blocks m_i, $1 \leq i \leq n$, a source node s and a sink node t. Second, for the vertical (horizontal) constraint graph, add a directed edge (v_i,v_j) if m_i is *below* (*left of*) m_j. Third, for each constraint graph, remove all edges that cannot be derived from other edges by transitivity.

Example: Floorplan to a Constraint-Graph Pair
Given: a floorplan with blocks *a-e* (right).

Task: generate the corresponding horizontal constraint graph (HCG) and vertical constraint graph (VCG).

Solution:

VCG and HCG: Create nodes *a-e* for blocks *a-e*. Create the source node *s* and sink node *t*.

VCG: Add a directed edge (v_i, v_j) if m_i is below m_j.

HCG: Add a directed edge (v_i, v_j) if m_i is left of m_j.

VCG and HCG: Remove all transitive (redundant) edges.

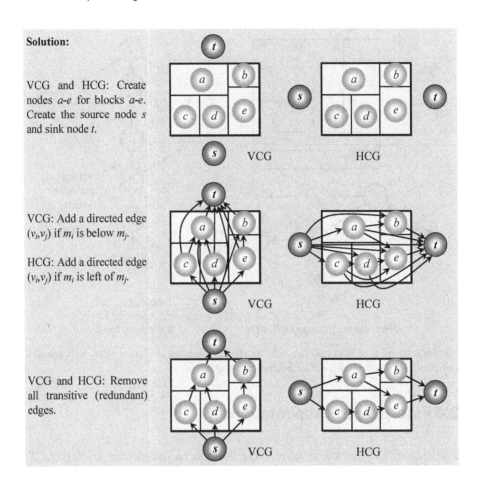

► **3.4.2 Floorplan to a Sequence Pair**

A sequence pair (S_+, S_-) encodes the same relations as the constraint-graph pair. Given two blocks a and b, in the sequence pair, if a comes before b in S_+, i.e., $<...a...b...>$, and a comes *before* b in S_-, i.e., $<...a...b...>$, then a is to the *left* of b. If a comes before b in S_+, i.e., $<...a...b...>$, and a comes *after* b in S_-, i.e., $<...b...a...>$, then a is *above* b. Conceptually, the constraint graphs are generated first, and then the block ordering rules are applied. However, the constraint graphs do not need to be created explicitly. Every pair of *non-overlapping* blocks is ordered either horizontally or vertically, based on the block locations. These ordering relations are encoded as constraints. When either constraint can be chosen, ties are broken, e.g., to always establish a horizontal constraint. Consider blocks a and b with locations (x_a, y_a) and (x_b, y_b) and dimensions (w_a, h_a) and (w_b, h_b), respectively.

if $((x_a + w_a \leq x_b)$ and not $(y_a + h_a \leq y_b$ or $y_b + h_b \leq y_a))$, then a is left of b

if $((y_b + h_b \leq y_a)$ and not $(x_a + w_a \leq x_b$ or $x_b + w_b \leq x_a))$, then a is above b

Example: A Floorplan to a Sequence Pair
Given: a floorplan with blocks *a-e* (right, same as in Sec. 3.4.1).
Task: generate the corresponding sequence pair.

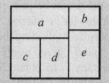

Solution:
Generate the VCG and HCG (example in Sec. 3.4.1). Evaluate each graph independently.

VCG: Blocks *c* and *d* are below block *a*.
Sequence pair so far: S_+: <...*a*...*c*...*d*...> S_-: <...*c*...*d*...*a*...>
VCG: Block *e* is below block *b*.
Sequence pair so far: S_+: <*acdbe*> S_-: <*cdaeb*>

HCG: Block *a* is left of blocks *b* and *e*.
Sequence pair so far: S_+: <*acdbe*> S_-: <*cdaeb*>
HCG: Block *c* is left of block *d*.
Sequence pair so far: S_+: <*acdbe*> S_-: <*cdaeb*>
HCG: Block *d* is left of block *e*.
Sequence pair: S_+: <*acdbe*> S_-: <*cdaeb*>

▶ 3.4.3 Sequence Pair to a Floorplan

Given a sequence pair, the following additional information is needed to generate a floorplan: (1) the origin of the floorplan, (2) the width and height of each block, and (3) the packing direction of the floorplan. The origin denotes the starting location at which the first block should be placed. The block dimensions are needed to calculate the relative displacements between blocks. The packing direction facilitates area minimization. Given different floorplan packing strategies, e.g., packing as left and downward as possible, or as right and upward as possible, different floorplans can be generated. For this section, assume that the packing direction is left and down.

Evaluating a sequence pair to generate a floorplan is closely related to finding the *weighted longest common subsequence* (*LCS*) of the two sequences [3.13]. That is, finding the *x*-coordinates of each block is equivalent to computing $LCS(S_+,S_-)$, and finding the *y*-coordinates of each block is equivalent to computing $LCS(S_+^R,S_-)$, where S_+^R is the reverse of S_+.

Sequence Pair Evaluation Algorithm
Input: sequence pair <S_+,S_->, widths (heights) of *n* blocks *widths[n]* (*heights[n]*)
Output: *x-* (*y-*) coordinates *x_coords* (*y_coords*), dimensions of floorplan $W \times H$
1.   for ($i = 1$ to n)
2.   *weights[i]* = *widths[i]*   // weights vector as block widths
3.   (*x_coords,W*) = LCS($S_+,S_-,weights$)   // x-coordinates, total width *W*
4.   for ($i = 1$ to n)
5.   *weights[i]* = *heights[i]*   // weights vector as block heights
6. $S_+^R[i] = S_+[n + 1 - i]$ // reverses S_+
7.   (*y_coords,H*) = LCS($S_+^R,S_-,weights$)   // y-coordinates, total height *H*

In the sequence pair evaluation algorithm, lines 1-2 initialize *weights* as the block widths for *n* blocks. Line 3 computes the *x*-coordinates of each block of the floorplan by finding the weighted longest common subsequence (*LCS* algorithm) of S_+ and S_-, given *weights*. Lines 4-5 re-initialize *weights* as the block heights. Line 6 reverses S_+ to obtain S_+^R. Line 7 computes the *y*-coordinates of each block of the floorplan by finding the weighted LCS of S_+^R and S_-, given *weights*.

Longest Common Subsequence (LCS) Algorithm
Input: sequences S_1 and S_2, weights of *n* elements *weights[n]*
Output: positions of each block *positions*, total span *L*

1.	**for** (*i* = 1 **to** *n*)	
2.	*block_order*[$S_2[i]$] = *i*	// index in S_2 of each block
3.	*lengths[i]* = 0	// initialize total span of all blocks
4.	**for** (*i* = 1 **to** *n*)	
5.	*block* = $S_1[i]$	// current block
6.	*index* = *block_order*[*block*]	// index in S_2 of current block
7.	*positions*[*block*] = *lengths*[*index*]	// compute position of block
8.	*t_span* = *positions*[*block*] + *weights*[*block*]	// find span of current block
9.	**for** (*j* = *index* **to** *n*)	
10.	**if** (*t_span* > *lengths[j]*)	
11.	*lengths[j]* = *t_span*	// current block span > previous
12.	**else break**	
13.	*L* = *lengths[n]*	// total span

The procedure $LCS(S_1,S_2,weights)$ computes the weighted LCS of two sequences S_1 and S_2 with weights *weights*. The vector *block_order* records the index in S_2 of each block (lines 1-2). Line 3 initializes *lengths*, the vector that stores the (maximum) span (representing the height or width) of each block, to 0. The variable *block* (line 4) corresponds to the current block in S_1 (line 5); *index* is for the index in S_2 of this current block (line 6). The position of *block* is then set to the first position not occupied by other blocks (line 7). That is, all blocks to the left of *block* are packed into the interval up to *lengths*[*block*], and *block* is positioned immediately afterward. Lines 9-12 update *lengths* to reflect the new total span of the floorplan – e.g., the last element *lengths[n]* stores the total span after the locations of all *n* blocks have been determined (line 13). Lines 10-11 update *lengths[j]* with (position of *block* + weight of *block*) if this exceeds the current span.

To find the *x*-coordinates of the floorplan, the LCS algorithm is called with $S_1 = S_+$, $S_2 = S_-$, and *weights* = *widths*. To find the *y*-coordinates, the LCS algorithm is called with $S_1 = S_+^R$, $S_2 = S_-$, and *weights* = *heights*.

Example: Sequence Pair to a Floorplan
Given: (1) sequence pair S_+: <*acdbe*> S_-: <*cdaeb*>, (2) packing direction: left and down, (3) floorplan origin (0,0), and (4) blocks *a-e* with their dimensions.

$A: (w_a,h_a) = (8,4)$ $B: (w_b,h_b) = (4,3)$
$C: (w_c,h_c) = (4,5)$ $D: (w_d,h_d) = (4,5)$
$E: (w_e,h_e) = (4,6)$

Task: generate the corresponding floorplan.

Solution:
widths[*a b c d e*] = [8 4 4 4 4]   *heights*[*a b c d e*] = [4 3 5 5 6]

Find *x*-coordinates.
$S_1 = S_+ = <acdbe>$, $S_2 = S_- = <cdaeb>$
weights[*a b c d e*] = *widths*[*a b c d e*] = [8 4 4 4 4]
block_order[*a b c d e*] = [3 5 1 2 4]
lengths = [0 0 0 0 0]

Iteration *i* = 1: *block* = *a*
index = *block_order*[*a*] = 3
positions[*a*] = *lengths*[*index*] = *lengths*[3] = 0
t_span = *positions*[*a*] + *weights*[*a*] = 0 + 8 = 8
Update *lengths* vector from *index* = 3 to *n* = 5: *lengths* = [0 0 **8 8 8**]

Iteration *i* = 2: *block* = *c*
index = *block_order*[*c*] = 1
positions[*c*] = *lengths*[*index*] = *lengths*[1] = 0
t_span = *positions*[*c*] + *weights*[*c*] = 0 + 4 = 4
Update *lengths* vector from *index* = 1 to *n* = 5: *lengths* = [**4 4 8 8 8**]

Iteration *i* = 3: *block* = *d*
index = *block_order*[*d*] = 2
positions[*d*] = *lengths*[*index*] = *lengths*[2] = 4
t_span = *positions*[*d*] + *weights*[*d*] = 4 + 4 = 8
Update *lengths* vector from *index* = 2 to *n* = 5: *lengths* = [4 **8 8 8 8**]

Iteration *i* = 4: *block* = *b*
index = *block_order*[*b*] = 5
positions[*b*] = *lengths*[*index*] = *lengths*[5] = 8
t_span = *positions*[*b*] + *weights*[*b*] = 8 + 4 = 12
Update *lengths* vector from *index* = 5 to *n* = 5: *lengths* = [4 8 8 8 **12**]

Iteration *i* = 5: *block* = *e*
index = *block_order*[*e*] = 4
positions[*e*] = *lengths*[*index*] = *lengths*[4] = 8
t_span = *positions*[*e*] + *weights*[*e*] = 8 + 4 = 12
Update *lengths* vector from *index* = 4 to *n* = 5: *lengths* = [4 8 8 **12 12**]

x-coordinates: *positions*[*a b c d e*] = [0 8 0 4 8], width of floorplan $W = lengths[5] = 12$.

Find *y*-coordinates.
$S_1 = S_+^R = <ebdca>$, $S_2 = S_- = <cdaeb>$
weights[*a b c d e*] = *heights*[*a b c d e*] = [4 3 5 5 6]
block_order[*a b c d e*] = [3 5 1 2 4]
lengths = [0 0 0 0 0]

Iteration i = 1: *block* = *e*
index = *block_order*[*e*] = 4
positions[*e*] = *lengths*[*index*] = *lengths*[4] = 0
t_span = *positions*[*e*] + *weights*[*e*] = 0 + 6 = 6
Update *lengths* vector from *index* = 4 to *n* = 5: *lengths* = [0 0 0 6 6]

Iteration i = 2: *block* = *b*
index = *block_order*[*b*] = 5
positions[*b*] = *lengths*[*index*] = *lengths*[5] = 6
t_span = *positions*[*b*] + *weights*[*b*] = 6 + 3 = 9
Update *lengths* vector from *index* = 5 to *n* = 5: *lengths* = [0 0 0 6 9]

Iteration i = 3: *block* = *d*
index = *block_order*[*d*] = 2
positions[*d*] = *lengths*[*index*] = *lengths*[2] = 0
t_span = *positions*[*d*] + *weights*[*d*] = 0 + 5 = 5
Update *lengths* vector from *index* = 2 to *n* = 5: *lengths* = [0 5 5 6 9]

Iteration i = 4: *block* = *c*
index = *block_order*[*c*] = 1
positions[*c*] = *lengths*[*index*] = *lengths*[1] = 0
t_span = *positions*[*c*] + *weights*[*c*] = 0 + 5 = 5
Update *lengths* vector from *index* = 1 to *n* = 5: *lengths* = [5 5 5 6 9]

Iteration i = 5: *block* = *a*
index = *block_order*[*a*] = 3
positions[*a*] = *lengths*[*index*] = *lengths*[3] = 5
t_span = *positions*[*a*] + *weights*[*a*] = 5 + 4 = 9
Update *lengths* vector from *index* = 3 to *n* = 5: *lengths* = [5 5 9 9 9]

y-coordinates: *positions*[*a b c d e*] = [5 6 0 0 0], height of floorplan H = *lengths*[5] = 9.

Floorplan size: W = 12 × H = 9
Coordinates of blocks *A-E*:
a (0,5)   *b* (8,6)   *c* (0,0)   *d* (4, 0)   *e* (8,0)

3.5 Floorplanning Algorithms

This section presents several algorithms used in floorplan optimization. Given a set of blocks, *floorplan sizing* determines the minimum area of the floorplan as well as the associated orientations and dimensions of each individual block. Techniques such as *cluster growth* and *simulated annealing* comprehend the netlist of interconnects between blocks and seek to (1) minimize the total length of interconnect, subject to an upper bound on the floorplan area, or (2) simultaneously optimize wirelength and area.

▶ 3.5.1 Floorplan Sizing

Floorplan sizing finds the dimensions of the minimum-area floorplan and corresponding dimensions of the individual blocks. The earliest algorithms, due to *Otten* [3.8] and *Stockmeyer* [3.11], make use of flexible dimensions for the individual blocks to find minimal top-level floorplan areas and shapes. A minimal top-level floorplan is selected, and the individual block shapes are chosen accordingly. Since the algorithm uses the shapes of both the individual blocks and the top-level floorplan, *shape functions* and *corner points* (limits) play a major role in determining an optimal floorplan.

Shape functions (shape curves) and **corner points.** Consider a block *block* with area *area(block)*. By definition, if a block has width w_{block} and height h_{block}, then

$$w_{block} \cdot h_{block} \geq area(block)$$

This relation can be rewritten in terms of width as a *shape function* (Fig. 3.6(a)):

$$h_{block}(w) = \frac{area(block)}{w}$$

That is, the shape function $h_{block}(w)$ states that for a given block width w, any block height $h \geq h_{block}(w)$ is legal. Shape functions can also include lower bounds on the block's width $LB(w_{block})$ and height $LB(h_{block})$. With such lower bounds, some (h,w) pairs are excluded (Fig. 3.6(b)).

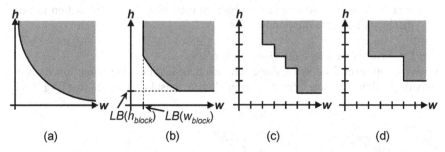

$LB(h_{block})$ $LB(w_{block})$

(a)   (b)   (c)   (d)

Fig. 3.6 Shape functions of blocks. The gray regions contains all (h,w) pairs that form a valid block [3.3]. (a) Shape function with no restrictions. (b) Shape function with minimum width $LB(w_{block})$ and height $LB(h_{block})$ restrictions for a block *block*. (c) Shape function with discrete (h,w) values. (d) Shape function of a possible hard library block, where its orientation can be more restricted.

Due to technology-dependent *design rules*, (h,w) pairs can be restricted to discrete values (Fig. 3.6(c)). Certain block libraries can impose even stronger restrictions. For instance, both the range of block dimensions, as well as the orientation of the block can be limited (Fig. 3.6(d)). Typically, blocks are also allowed to be reflected about the *x*- or *y*-axis.

The discrete dimensions of the block, sometimes called *outer* nodes (with respect to the feasible region of (h,w) pairs), can be thought of as non-dominated *corner points* that limit the shape function (Fig. 3.7). One academic floorplanner, *DeFer*, extends this calculus of shape functions to simultaneously optimize the shapes and locations of blocks in slicing floorplans [3.16].

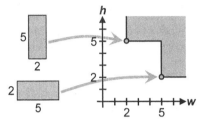

Fig. 3.7 A 2 × 5 rotatable library block that has corner points $w = 2$ and $w = 5$ in its shape function.

Minimum-area floorplan. This algorithm finds the minimum floorplan area for a given slicing floorplan in polynomial time. For non-slicing floorplans, the problem is *NP-hard* [3.3].

Floorplan sizing consists of three major steps.

1. Construct the shape functions of all individual blocks.
2. Determine the shape function of the top-level floorplan from the shape functions of the individual blocks using a bottom-up strategy. That is, start with the lowest-level blocks and perform horizontal and vertical composition until the optimal size and shape of the top-level floorplan are determined.
3. From the corner point that corresponds to the minimum top-level floorplan area, trace in a top-down fashion back to each block's shape function to find that block's dimensions and location.

Step 1: Construct the shape functions of the blocks
Since the top-level floorplan's shape function depends upon the shape functions of individual blocks, the shape functions of all blocks must be identified first (Fig. 3.8).

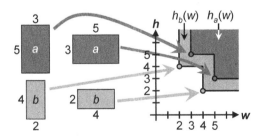

Fig. 3.8 Shape functions of two library blocks a (5×3) and b (4×2). The shape functions $h_a(w)$ and $h_b(w)$ show the feasible height-width combinations of the blocks.

Step 2: Determine the shape function of the top-level floorplan
The shape function of the top-level floorplan is derived from the individual blocks' shape functions. Two different kinds of *composition – vertical* and *horizontal –* can yield different results.

In Fig. 3.9, block *a* is *vertically* aligned with block *b*. Let the shape function of block *a* be $h_a(w)$, and let the shape function of block *b* be $h_b(w)$. Then, the shape function of the top-level floorplan *F* is

$$h_F(w) = h_a(w) + h_b(w)$$

The height of *F* is determined by h_F, adding $h_a(w)$ and $h_b(w)$ for every corner point. The width of *F* is found as $w_F = \max(w_a, w_b)$.

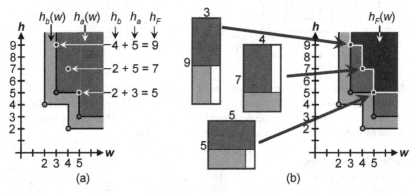

Fig. 3.9 Vertical composition of two library blocks *a* (5 × 3) and *b* (4 × 2) superimposed for vertical composition. (a) To find the smallest bounding box for the top-level floorplan $h_F(w)$ from the functions $h_a(w)$ and $h_b(w)$, the block heights of the respective corner points are added. In this example, $w = 3$ and $w = 5$ (from $h_a(w)$) are combined with $w = 4$ (from $h_b(w)$). Note that $w = 2$ (from $h_b(w)$) is ignored, since the width of the top-level floorplan cannot be smaller than 3, which is the width of block *a*. (b) The potential floorplans of *F* are the corner points of $h_F(w)$.

In Fig. 3.10, block *a* is *horizontally* aligned with block *b*. The width of *F* is determined by $w_F(h)$, adding $w_a(h)$ and $w_b(h)$ for every corner point. The height of *F* is $h_F = \max(h_a, h_b)$.

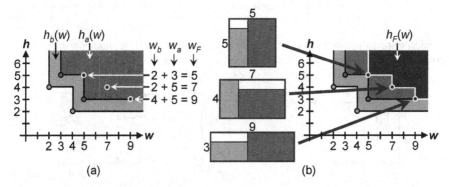

Fig. 3.10 Shape functions of two library blocks *a* (5 × 3) and *b* (4 × 2) superimposed for horizontal composition. (a) Finding $h_F(w)$. (b) The potential floorplans of *F* are the corner points of $h_F(w)$.

Step 3: Find the floorplan and individual blocks' dimensions and locations
Once the shape function of the top-level floorplan has been determined, the minimum floorplan area is computed. All minimum-area floorplans are always on corner points of the shape function (Fig. 3.11). After finding the corner point for the minimum area, each individual block's dimensions and relative location are found by backtracing from the floorplan's shape function to the block's shape function.

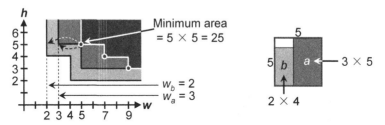

Fig. 3.11 Finding the shapes of the individual blocks from the same top-level floorplan example (Fig. 3.9 and Fig. 3.10) with library blocks a (5 × 3) and b (2 × 4). The minimum-area floorplan F has $w_F = 5$ and $h_F = 5$ (left). The dimensions and locations of the individual blocks a and b (right) are derived by tracing back to the blocks' respective shape functions.

Example: Floorplan Sizing
Given: two blocks a and b.
$a: w_a = 1, h_a = 3$ or $w_a = 3, h_a = 1$
$b: w_b = 2, h_b = 2$ or $w_b = 4, h_b = 1$

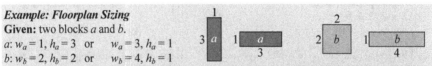

Task: find the minimum-area floorplan F using both horizontal and vertical composition, and its corresponding slicing tree.

Solution:
Construct the shape functions $h_a(w)$ and $h_b(w)$ of blocks a (left) and b (right).

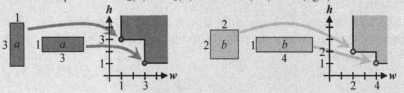

Vertical composition: determine the shape function $h_F(w)$ of F and the minimum-area corner point. The minimum area of F is 8, with dimensions $w_F = 4$ and $h_F = 2$.

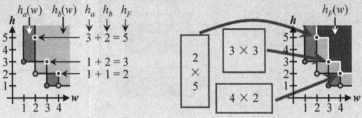

Find the dimensions and locations of blocks a and b.

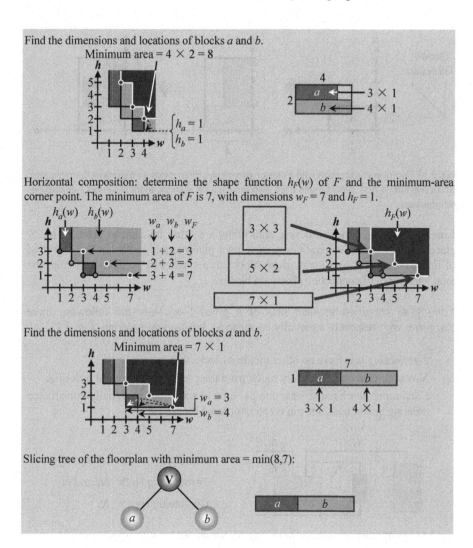

Horizontal composition: determine the shape function $h_F(w)$ of F and the minimum-area corner point. The minimum area of F is 7, with dimensions $w_F = 7$ and $h_F = 1$.

Find the dimensions and locations of blocks a and b.

Slicing tree of the floorplan with minimum area $= \min(8,7)$:

▶ 3.5.2 Cluster Growth

In a cluster growth-based method, the floorplan is constructed by iteratively adding blocks until all blocks have been assigned (Fig. 3.12). An initial block is chosen and placed in the lower-left (or any other) corner. Subsequent blocks are added, one at a time, and merged either horizontally, vertically, or diagonally with the cluster. The location and orientation of the next block depends on the current shape of the cluster; it will be placed to best accommodate the objective function of the floorplan. In contrast to the floorplan-sizing algorithm, only the different orientations of the individual blocks are taken into account. Methods that directly and simultaneously optimize both the shapes of each block and the floorplan are not currently known. The order of the blocks is typically chosen by a linear-ordering algorithm.

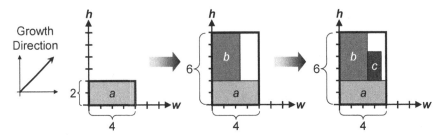

Fig. 3.12 Floorplan construction with a cluster-growth algorithm based on minimum floorplan area. Block *a* is placed first. Blocks *b* and *c* are placed so that the increase to the floorplan's dimensions is minimum.

Linear ordering. Linear-ordering algorithms are often invoked to produce initial placement solutions for iterative-improvement placement algorithms. The objective of linear ordering is to arrange given blocks in a single row so as to minimum the total wirelength of connections between blocks.

Kang [3.4] classified incident nets of a given block into the following three categories with respect to a partially constructed, left-to-right ordering (Fig. 3.13).

- *Terminating nets* have no other incident blocks that are unplaced.
- *New nets* have no pins on any block from the partially-constructed ordering.
- *Continuing nets* have at least one pin on a block from the partially-constructed ordering and at least one pin on an unordered block.

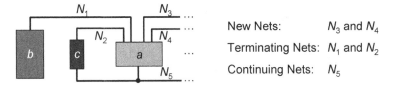

Fig. 3.13 Classification of nets based on the partially-constructed linear ordering.

The ordering of a particular block *m* directly depends on the types of nets attached to *m*. Specifically, the block that completes the greatest number of "unfinished" nets should be placed first. In other words, the block with the highest difference between the number of terminating and new nets is chosen as the next block in the sequence.

The linear-ordering algorithm starts by choosing an initial block. This can be done either arbitrarily or based on the number of connections to the other blocks (line 1). During every iteration, the gain *gain* for each block $m \in M$ is calculated (lines 5-8), where *gain* is the difference between the number of terminating nets and new nets of *m*. The blocks with the maximum gain are selected (line 9). In case there are multiple blocks that have the maximum gain value, the blocks with the most terminating nets are selected (lines 10-11). If there are still multiple blocks, the blocks with the most continuing nets are selected (lines 12-13). If there are still multiple blocks, the blocks with the lowest connectivity are selected (lines 14-15). If there are still multiple blocks, a block is selected arbitrarily (lines 16-17). The selected block is then added to *order* and removed from *M* (lines 18-19).

Linear Ordering Algorithm [3.4]
Input: set of all blocks *M*
Output: ordering of blocks *order*
1.   *seed* = starting block
2.   *order* = *seed*
3.   REMOVE(*M*,*seed*)   // remove *seed* from *M*
4.   **while** (|*M*| > 0)   // while *M* is not empty
5.   **foreach** (block *m* ∈ *M*)
6.   *term_nets*[*m*] = number of terminating nets incident to *m*
7.   *new_nets*[*m*] = number of new nets incident to *m*
8.   *gain*[*m*] = *term_nets*[*m*] − *new_nets*[*m*]
9.   *M'* = block(s) that have maximum *gain*[*m*]
10.   **if** (|*M'*| > 1)   // multiple blocks
11.   *M'* = block(s) with the most terminating nets
12.   **if** (|*M'*| > 1)   // multiple blocks
13.   *M'* = block(s) with the most continuing nets
14.   **if** (|*M'*| > 1)   // multiple blocks
15.   *M'* = block(s) with the fewest connected nets
16.   **if** (|*M'*| > 1)   // multiple blocks
17.   *M'* = arbitrary block in *M*
18.   ADD(*order*,*M'*)   // add *M'* to *order*
19.   REMOVE(*M*,*M'*)   // remove *M'* from *M*

Example: Linear Ordering Algorithm
Given: (1) netlist with five blocks *a-e*,
(2) starting block *a*, and (3) six nets N_1-N_6.
  $N_1 = (a,b)$ $N_2 = (a,d)$ $N_3 = (a,c,e)$
  $N_4 = (b,d)$ $N_5 = (c,d,e)$ $N_6 = (d,e)$

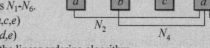

Task: find a linear ordering using the linear ordering algorithm.

Solution:

Iteration #	Block	New Nets	Terminating Nets	*gain*	Continuing Nets
0	*a*	N_1,N_2,N_3	--	-3	--
1	*b*	N_4	N_1	0	--
	c	N_5	--	-1	N_3
	d	N_4,N_5,N_6	N_2	-2	--
	e	N_5,N_6	--	-2	N_3
2	*c*	N_5	--	-1	N_3
	d	N_5,N_6	N_2,N_4	0	--
	e	N_5,N_6	--	-2	N_3
3	*c*	--	--	0	N_3,N_5
	e	--	N_6	1	N_3,N_5
4	*c*	--	N_3,N_5	2	--

For each iteration, **bold font** denotes the block with maximum gain.
Iteration 0: set block *a* as the first block in the ordering.
Iteration 1: block *b* has maximum gain. Set as the second block in the ordering.
Iteration 2: block *d* has maximum gain. Set as the third block in the ordering.
Iteration 3: block *e* has maximum gain. Set as the fourth block in the ordering.
Iteration 4: set *c* as the fifth (last) block in the ordering.

The linear ordering that heuristically
minimizes total net cost is <*a b d e c*>.

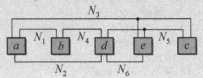

Cluster growth. In the cluster-growth algorithm, the blocks are ordered using the linear ordering algorithm (line 2). For each block *curr_block* (line 4), the algorithm finds a location such that the floorplan grows evenly – toward the upper and right sides – while satisfying criteria such as the shape constraint of the top-level floorplan (line 5). Other typical criteria include total wirelength or the amount of *deadspace* within the floorplan. This algorithm is similar to the *Tetris* algorithm used for cell legalization (Sec. 4.4).

Cluster-Growth Algorithm
Input: set of all blocks *M*, cost function *C*
Output: optimized floorplan *F* based on *C*
1. $F = \emptyset$
2.   *order* = LINEAR_ORDERING(*M*)   // generate linear ordering
3.   **for** (*i* = 1 **to** |*order*|)
4.   *curr_block* = *order*[*i*]
5.   ADD_TO_FLOORPLAN(*F*,*curr_block*,*C*)   // find location and orientation
                                                    // of *curr_block* that causes
                                                    // smallest increase based on
                                                    // *C* while obeying constraints

The example below illustrates the cluster-growth algorithm. It uses the same linear ordering found in the previous example. The objective is to find the smallest floorplan, i.e., to minimize the area of the top-level floorplan. Though it produces mediocre solutions, the cluster-growth algorithm is fast and easy to implement; therefore, it is often used to find initial floorplan solutions for iterative algorithms such as *simulated annealing*.

Example: Floorplan Construction by Cluster Growth
Given: (1) blocks *a-e* and (2) linear ordering <*a b d e c*>.
a: $w_a = 2, h_a = 3$ or $w_a = 3, h_a = 2$
b: $w_b = 2, h_b = 1$ or $w_b = 1, h_b = 2$
c: $w_c = 2, h_c = 4$ or $w_c = 4, h_c = 2$ Growth Direction:
d: $w_d = 3, h_d = 3$
e: $w_e = 6, h_e = 1$ or $w_e = 1, h_e = 6$
Task: find a floorplan with minimum global bounding box area.

Solution: (multiple solutions possible; one is illustrated below)
Block a: place in lower-left corner with $w_a = 2$ and $h_a = 3$.

Current bounding box area $= 2 \times 3 = 6$.

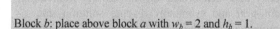

Block b: place above block a with $w_b = 2$ and $h_b = 1$.

Current bounding box area $= 2 \times 4 = 8$.

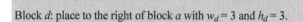

Block d: place to the right of block a with $w_d = 3$ and $h_d = 3$.

Current bounding box area $= 5 \times 4 = 20$.

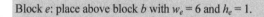

Block e: place above block b with $w_e = 6$ and $h_e = 1$.

Current bounding box area $= 6 \times 5 = 30$.

Block c: place to the right of block d with $w_c = 2$ and $h_c = 4$.

Current bounding box area $= 7 \times 5 = 35$.

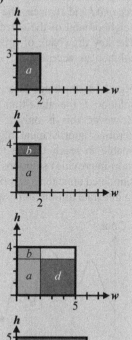

▶ 3.5.3 Simulated Annealing

Simulated annealing (*SA*) algorithms are iterative in nature – they begin with an initial (arbitrary) solution and seek to incrementally improve the objective function. During each iteration, a local *neighborhood* of the current solution is considered. A new candidate solution is formed by a small perturbation to the current solution within its neighborhood.

Unlike *greedy algorithms*, SA algorithms do not always reject candidate solutions with higher cost. That is, in a greedy algorithm, if the new solution is better, e.g., has lower cost (assuming a minimization objective), than the current solution, it is *accepted* and replaces the current solution. If no better solution exists in the local neighborhood of the current solution, the algorithm has reached a *local minimum*. The key drawback of greedy algorithms is only beneficial changes to the current solution are accepted.

Fig. 3.14 illustrates how the greedy approach breaks down. Starting from initial solution *I*, the algorithm goes *downhill* and eventually reaches solution state *L*. However, this is only a *local minimum*, not the *global minimum G*. A greedy iterative improvement algorithm, unless given a fortuitous initial state, will be unable to reach *G*. On the other hand, iterative approaches that accept inferior (non-improving) solutions can *hill-climb* away from *L* and potentially reach *G*. The simulated annealing algorithm is one of the most successful optimization strategies to integrate hill-climbing with iterative improvement.

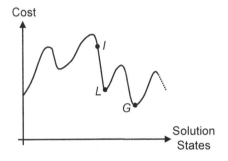

Fig. 3.14 For a minimization problem, an initial solution *I*, along with local optimum (*L*) and global optimum (*G*) solutions. In the neighborhood structure shown, the cost function has several local minima, as is the case with most (intractable) layout problems.

Principle of simulated annealing. In materials science, *annealing* refers to the controlled cooling of high-temperature materials to modify their properties. The goal of annealing is to alter the atomic structure of the material and reach a *minimum-energy configuration*. For instance, the atoms of a high-temperature metal are in high-energy, disordered states (chaos), while the atoms of a low-temperature metal are in low-energy, ordered states (crystalline structures). However, the same metal may experience atomic configurations that are brittle or mechanically hard, depending on the size of the individual crystals. When the high-temperature metal is cooled, it goes from a highly randomized state to a more structured state. The rate at which the metal is cooled will drastically affect the final structure. Moreover, the way in which the atoms settle is *probabilistic* in nature. In practice, a slower cooling process implies a greater chance that the atoms will settle as a perfect lattice, forming a minimum-energy configuration.

The cooling process occurs in steps. At each step of the *cooling schedule*, the temperature is held constant for a specified amount of time. This allows the atoms to gradually cool down and stabilize, i.e., reach a *thermodynamic equilibrium*, at each given temperature. Though the atoms have the ability to move across large distances and create new higher-energy states, the probability of such drastic changes in this configuration decreases with temperature.

As the cooling process continues, the atoms eventually will settle in a local, and possibly global, minimum-energy configuration. Both the rate and step size of the temperature decrease will affect how the atoms will settle. If the rate is sufficiently slow and the increment is sufficiently small, the atoms will settle, with high probability, at a global minimum. On the other hand, if cooling is too fast or the increment is too large, then the atoms are less likely to attain the global minimum-energy configuration, and instead will settle in a local minimum instead.

Annealing-based optimization. The principle of annealing can be applied to solve combinatorial optimization problems. In the context of minimization, finding the lowest-cost solution in an optimization problem is analogous to finding a minimum-energy state of a material. Thus, *simulated annealing* algorithms take a "chaotic" (higher-cost) solution and emulate physical annealing to produce a "structured" (lower-cost) solution.

The simulated annealing algorithm generates an initial solution and evaluates its cost. At each step, the algorithm generates a new solution by performing a *random walk* in the solution by applying a small perturbation (change in structure). This new solution is then *accepted* or *rejected* based on a *temperature* parameter T. When T is high (low), the algorithm has a higher (lower) chance of accepting a solution with higher cost. Analogous to physical annealing, the algorithm slowly decreases T, which correspondingly decreases the probability of accepting an inferior, higher-cost solution. One method for probabilistically accepting moves is based on the *Boltzmann acceptance criterion*, where the new solution is acceptance if

$$e^{-\frac{cost(curr\_sol)-cost(next\_sol)}{T}} > r$$

Here, *curr_sol* is the current solution, *next_sol* is the new solution after a perturbation, T is the current temperature, and r is a random number between $[0,1)$ based on a uniform distribution. For a minimization problem, the final solution will be in a *valley*; for a maximization problem, it will be at a *peak*.

The rate of temperature decrease is extremely important – it (1) must enable sufficient high-temperature exploration of the solution space at the beginning, while (2) allowing enough time at low temperatures to have sufficient probability of settling at a near-optimal solution. Just as slow cooling of high-temperature metal has a high probability of finding a globally optimal, energy-minimal crystal lattice, a simulated annealing algorithm with a sufficiently slow cooling schedule has high probability of finding a high-quality solution for a given optimization problem [3.5].

The simulated annealing algorithm is *stochastic* by nature – two runs usually yield two different results. The difference in quality stems from probabilistic decisions such as generation of new, perturbed solutions (e.g., by a cell swap), and the acceptance or rejection of moves.

Algorithm. The algorithm begins with an initial (arbitrary) solution *curr_sol* (lines 3-4), and generates a new solution by perturbing *curr_sol* (lines 8-9). The resulting new cost *trial_cost* is computed (line 10), and compared with the current cost *curr_cost* (line 11). If the new cost is better, i.e., $\Delta cost < 0$, the change of solution is accepted (lines 12-14). Otherwise, the change may still be probabilistically accepted. A random number $0 \leq r < 1$ is generated (line 16), and if r is smaller than $e^{-\Delta cost/T}$, the change is accepted. Otherwise, the change is rejected (lines 17-19).

To perform maximization, reverse the order of operands for the subtraction in line 11. In the context of floorplanning, the function TRY_MOVE (line 9) can be replaced with operations such as moving a single block or swapping two blocks.

Simulated Annealing Algorithm
Input: initial solution *init_sol*
Output: optimized new solution *curr_sol*

1.	$T = T_0$	// initialization
2.	$i = 0$	
3.	*curr_sol* = *init_sol*	
4.	*curr_cost* = COST(*curr_sol*)	
5.	**while** $(T > T_{min})$	
6.	**while** (stopping criterion is not met)	
7.	$i = i + 1$	
8.	(a_i, b_i) = SELECT_PAIR(*curr_sol*)	// select two objects to perturb
9.	*trial_sol* = TRY_MOVE(a_i, b_i)	// try small local change
10.	*trial_cost* = COST(*trial_sol*)	
11.	$\Delta cost$ = *trial_cost* − *curr_cost*	
12.	**if** $(\Delta cost < 0)$	// if there is improvement,
13.	*curr_cost* = *trial_cost*	// update the cost and
14.	*curr_sol* = MOVE(a_i, b_i)	// execute the move
15.	**else**	
16.	r = RANDOM(0,1)	// random number [0,1)
17.	**if** $(r < e^{-\Delta cost/T})$	// if it meets threshold,
18.	*curr_cost* = *trial_cost*	// update the cost and
19.	*curr_sol* = MOVE(a_i, b_i)	// execute the move
20.	$T = \alpha \cdot T$	// $0 < \alpha < 1$, T reduction

The probability of the exchange depends on both cost and temperature, and a larger (worse) cost difference implies a smaller chance that the change is accepted. At high temperatures, i.e., $T \to \infty$, $-\Delta cost / T \approx 0$ and the probability of accepting a change approaches $e^0 = 1$. Again, this leads to frequent acceptance of inferior solutions, which helps the algorithm escape "low-quality" regions of the solution space. On the other hand, at low temperatures, i.e., $T \to 0$, $-\Delta cost / T \approx \infty$ and the probability of accepting a change approaches $e^{-\infty} = 1/e^{\infty} = 0$. The initial temperature (T_0) and the degradation rate (α) are usually set empirically. In general, better results are produced with a high initial temperature and a slow rate of cooling, at the cost of increased runtime.

Simulated annealing-based floorplanning. The first simulated annealing algorithm for floorplanning was proposed in 1984 by *R. Otten* and *L. van Ginneken* [3.9]. Since then, simulated annealing has become one of the most common iterative methods used in floorplanning.

In the *direct approach*, SA is applied directly to the physical layout, using the actual coordinates, sizes, and shapes of the blocks. However, finding a fully legal solution – a floorplan with no block overlaps – is difficult. Thus, intermediate solutions are allowed to have overlapping blocks, and a penalty function is incorporated to encourage legal solutions. The final produced solution, though, must be completely *legal* (see [3.10] for further reading).

In the *indirect approach*, simulated annealing is applied to an abstraction of the physical layout. Abstract representations capture the floorplan using trees or constraint graphs. A final mapping is also required to generate the floorplan from the abstract representation. One advantage of this process over the direct approach is that all intermediate solutions are overlap-free.

For further reading on simulated annealing-based floorplanning, see [3.1], [3.3], [3.14] and [3.15].

▶ 3.5.4 Integrated Floorplanning Algorithms

Analytic techniques map the floorplanning problem to a set of equations where the variables represent block locations. These equations describe boundary conditions, attempt to prevent block overlap, and capture other relations between blocks. In addition, an objective function quantifies the important parameters of the floorplan.

One well-known analytic method is *mixed integer-linear programming* (*MILP*), where the location variables are integers. This technique does not allow for overlaps and seeks globally optimal solutions. However, it is limited due to its computational complexity. For a problem size of 100 blocks, the integer program can have over 10,000 variables and over 20,000 equations. Thus, MILP is usable only for small (10 or fewer blocks) instances.

A faster alternative that offers some compromises is to use a linear programming (LP) relaxation. Compared to MILP, the LP formulation does not limit the locations to be integers. However, LP can be used for larger problem instances.

For further discussion of floorplanning with analytic methods, see [3.1]. A technique for *floorplan repair* (legalization) is described in [3.7].

3.6 Pin Assignment

Given the large geometric sizes of blocks during floorplanning, the terminal locations of nets connecting these blocks are very important. *I/O pins* (net terminals) and their locations are usually on the periphery of a block to reduce interconnect length. However, the best locations depend on the relative placement of the blocks.

Problem formulation. *During pin assignment, all nets (signals) are assigned to unique pin locations such that the overall design performance is optimized. Common optimization goals include maximizing routability and minimizing electrical parasitics both inside and outside of the block.*

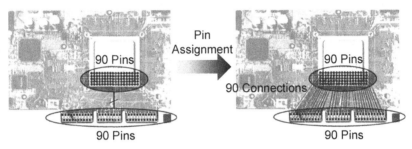

Fig. 3.15 The pin assignment process. Here, each of the 90 I/O pins from the chip is assigned a specific I/O pin on a printed circuit board. Each pin pair is then connected by a route.

The goal of *external pin assignment* is to connect each incoming or outgoing signal to a unique I/O pin. Once the necessary nets have each been assigned a unique pin, they must be connected such that wirelength and electrical parasitics, e.g., coupling or reduced signal integrity, are minimized. For instance, Fig. 3.15 shows 90 pins on the microprocessor chip, each of which must be connected to an I/O pad at the next hierarchy level. After pin assignment, each pin from the chip has a connection to a unique pin on the external device, connected with short routes.

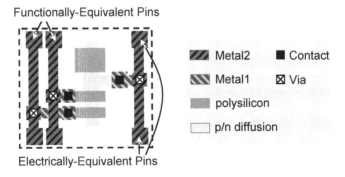

Fig. 3.16 Functionally-equivalent input pins and electrically-equivalent output pins for a simplified example of an nMOS NAND gate.

Alternatively, pin assignment can be used to connect cell pins that are *functionally* or *electrically equivalent*, such as during standard-cell placement (Chap. 4). Two pins are functionally equivalent if swapping them does not affect the design's logic and electrically equivalent (equipotential) if they are connected (Fig. 3.16).

The main objective of *internal pin assignment* for cells is to reduce congestion and interconnect length between the cells (Fig. 3.17). Pin assignment techniques discussed below apply to both chip planning and placement (Chap. 4).

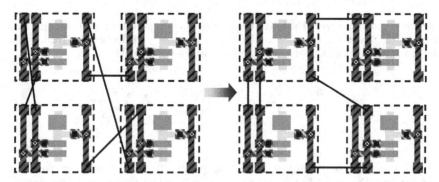

Fig. 3.17 Pin assignment using the example given in Fig. 3.16. The assignment aims to minimize the total connection length by exploiting functionally and electrically equivalent pins.

Pin assignment using concentric circles. The objective of this algorithm is to establish connections between a block and all its related pins in other blocks such that net crossings are minimized. The following simple algorithm [3.6], introduced in 1972, assumes that all *outer pins* (pins outside of the current block) have fixed locations. All *inner pins* (pins in the current block) will be assigned locations based on the locations of the electrically-equivalent outer pins. The algorithm uses two concentric circles – an *inner circle* for the pins of the block under consideration, and an *outer circle* for the pins in the other blocks. The goal is to assign legal pin locations to both circles such that there is no net overlap.

Example: Pin Assignment Using Concentric Circles (Including Algorithm)
Given: (1) set of pins on the block (black) and (2) set of pins on the chip (white) to which the block pins must connect.
Task: perform pin assignment using concentric circles such that each pin on the block is connected to exactly one pin on the chip, and vice-versa (i.e., a 1-to-1 mapping).

Solution:
Determine the circles. The two circles are drawn such that (1) all pins that belong to the block (black) are outside of the inner circle but within the outer circle and (2) all external pins (white) are outside of the outer circle.

Determine the points. For each point, draw a line from that point to the center of the circles (left). Move each outer (white) point to its projection on the outer circle, and move each inner (black) point to its projection on the inner circle (right).

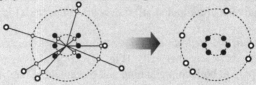

Determine initial mapping. The initial mapping is an assignment from each outer pin to a corresponding inner pin. Choose a starting point and assign it arbitrarily (left). Then, assign the remaining points in clockwise or counter-clockwise direction (right).

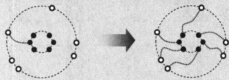

Optimize the mapping. Repeat the mapping process for other outer-inner point-pair combinations. That is, for the same starting point on the outer circle, assign a different point on the inner circle and map the remaining points. Do this until all point-pair combinations have been considered. The best mapping is the one with the shortest Euclidean distance. For the problem instance, a possible mapping is shown on the left, the best mapping is shown in the center, and the final pin assignment is shown on the right.

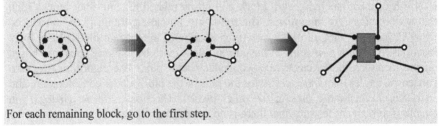

For each remaining block, go to the first step.

Topological pin assignment. In 1984, *H. N. Brady* improved the concentric-circle pin assignment algorithm by taking into account external block positions and multi-pin nets (connected to more than two pins) [3.2]. Specifically, this enabled pin assignment when external pins are behind other blocks or obstacles.

Fig. 3.18(a) shows an example assigning pins from the main component *m* to an external block *b*. A *midpoint line* $l_{m\sim b}$ is drawn from the center of *m* through the midpoint of *b*. In Fig. 3.18(b), the pins of *b* are "unwrapped" and expanded as a line *l'* at the *dividing point d* – the farther point on $l_{m\sim b}$ that intersections *b*. The pins are then projected onto the outer circle (from the original concentric- circle algorithm).

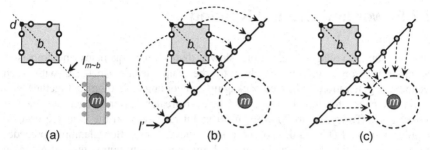

Fig. 3.18 Pin assignment from the main component m to an external block b. (a) The main component m is collapsed as a single point. A midpoint line $l_{m\sim b}$ is drawn from m through the center of b. A dividing point d is formed at the farther point where $l_{m\sim b}$ intersects b. (b) From d, project b's pins onto l'. (c) The pins are then projected onto the outer circle of m [3.2].

This improvement also allows several blocks to be considered simultaneously. Let the set of all external blocks be denoted as B, and let the main component be denoted as m. A midpoint line $l_{m\sim b}$ is drawn from m through the midpoint of every external block $b \in B$.

A dividing point on block b is formed at the *farther* point where $l_{m\sim b}$ intersects b, and at the *closer* point where each midpoint line $l_{m\sim b'}$ intersect b, where $b' \in B$ is located between m and b. Based on these dividing points, the pins are separated and extended accordingly (Fig. 3.19).

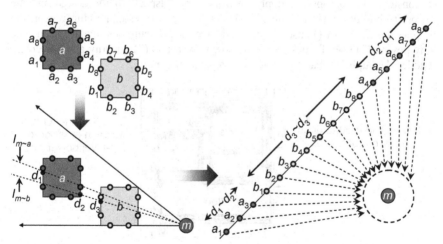

Fig. 3.19 Pin assignment on blocks a and b [3.2]. On block a, consider the midpoint lines $l_{m\sim a}$ and $l_{m\sim b}$. The dividing point d_1 is formed because it is the farther point on $l_{m\sim a}$; the dividing point d_2 is formed because it is the closer point on $l_{m\sim b}$. On block b, consider only $l_{m\sim b}$. The dividing point d_3 is formed because it is the farther point on $l_{m\sim b}$. Using d_1-d_3, the pins are "unwrapped" accordingly when projected onto m's outer circle.

3.7 Power and Ground Routing

On-chip supply voltages scale more slowly than chip frequencies and transistor counts. Therefore, currents supplied to the chip steadily increase with each technology generation. Improved packaging and cooling technologies, together with market demands for functionality, lead to ever-greater power budgets and denser power grids. Today, up to 20-40% of all metal resources on the chip are used to supply *power* (*VDD*) and *ground* (*GND*) nets. Since floorplanning precedes place-and-route, i.e., block- and chip-level implementation, power-ground planning has become an essential part of the modern chip planning process.

Chip planning determines not only the layout of the power-ground distribution network, but also the placement of *supply I/O pads* (with wire-bond packaging) or *bumps* (with flip-chip packaging). The pads or bumps are preferentially located in or near high-activity regions of the chip to minimize the $V = IR$ voltage drop.[2] In general, the power planning process is highly iterative, including (1) early simulation of major power dissipation components, (2) early quantification of chip power, (3) analyses of total chip power and maximum power density, (4) analyses of total chip power fluctuations, (5) analyses of inherent and added fluctuations due to clock gating, and (6) early power distribution analysis: average, maximum and multi-cycle fluctuations.

To construct an appropriate supply network, many aspects of the design and the process technology must be considered. For example, to estimate chip power, the designer must plan for (1) use of low-V_{th} devices and dynamic circuits that consume more power, (2) use of clock gating for low power, and (3) quantity and placement of added decoupling capacitors that mitigate switching noise.

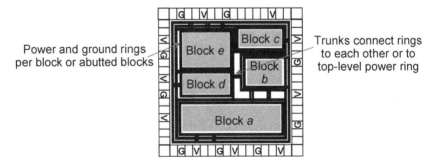

Fig. 3.20 Custom style of power-ground distribution for a chip floorplan.

---

[2] A supply I/O pad can deliver tens of milliamperes of current, while a supply bump can deliver hundreds of milliamperes, i.e., an order of magnitude more current. As an example of how much die resource goes to power-ground distribution – the Intel Pentium 4 microprocessor chip uses 423 bumps, of which 223 are for delivery of *VDD* and *GND*.

This section discusses the physical design of *power-ground distribution networks*. Fig. 3.20 illustrates conceptually how a floorplan in a *custom* design approach might associate supply rings with each block, for later connection to a chip-level power distribution plan such as those discussed below.

▶ **3.7.1 Design of a Power-Ground Distribution Network**

The *supply nets*, *VDD* and *GND*, connect each cell in the design to a power source. As each cell must have both *VDD* and *GND* connections, the supply nets (1) are large, (2) span across the entire chip, and (3) are routed first before any signal routing. *Core* supply nets are distinguished *I/O* supply nets, which are typically at a higher voltage. In many applications, one core power net and one core ground net are sufficient. Some ICs, such as mixed-signal or low-power (supply-gated or multiple voltage level) designs, can have multiple power and ground nets.

Routing of supply nets is different from routing of signals. Power and ground nets should have dedicated metal layers to avoid consuming signal routing resources. In addition, supply nets prefer thick metal layers – typically, the top two layers in the back-end-of-line process – due to their low resistance. When the power-ground network traverses multiple layers, there must be sufficient *vias* to carry current while avoiding *electromigration* and other *reliability* issues.

Since supply nets have high current loads, they are often much wider than standard signal routes. The widths of the individual wire segments may be tailored to accommodate their respective estimated branch currents. For logic gates to have correct timing performance, the net segment width must be chosen to keep the voltage drop, $V = IR$, within a specified tolerance, e.g., 5% of *VDD*. Wider segments have lower resistance, and hence lower voltage drop.[3]

There are two approaches to the physical design of power-ground distribution – the *planar* approach, which is used primarily in analog or custom blocks, and the *mesh* approach, which is predominant in digital ICs.

▶ **3.7.2 Planar Routing**

Power supply nets can be laid out using *planar routing* when (1) only two supply nets are present in the design, and (2) a cell needs a connection to both supply nets. Planar routing separates the two supply regions by a *Hamiltonian path* that connects all the cells, such that each supply net can be attached either to the *left* or *right* of every cell. The Hamiltonian path allows both supply nets to be routed across the layout – one to the left and one to the right of the path– with no *conflicts* (Fig. 3.21).

---

[3] Some design manuals will refer to an *IR* drop limit of 10% of *VDD*. This means that the supply can *drop* (*droop*) by 5% of *VDD* and the ground can *bounce* by 5% as well, resulting in a worst-case of 10% supply reduction.

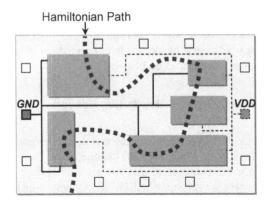

Fig. 3.21 Given a Hamiltonian path that connects all cells, each supply net has its own "uninterrupted" path to each cell, and hence both supply nets can be routed on one layer.

Routing the power and ground nets in this planar fashion can be accomplished with the following three steps.

Step 1: Planarize the topology of the nets

As both power and ground nets must be routed on one layer, the design should be split using the Hamiltonian path. In Fig. 3.22(a), the two nets start from the left and right sides. Both nets grow in a tree-like fashion, without any conflict (overlap), and separated by the Hamiltonian path. The exact routes will depend on the pin locations. Finally, cells are connected wherever a pin is encountered.

Step 2: Layer assignment

Net segments are assigned to appropriate routing layers based on routability, the resistance and capacitance properties of each available layer, and design rule information.

Step 3: Determining the widths of the net segments

The width of each segment (branch) depends on the maximum current flow. That is, a segment's width is determined from the sum of the currents from all the cells to which it connects in accordance with *Kirchhoff's Current Law* (*KCL*). When dealing with large currents, designers often extend the "width" of the "planar" route in the vertical dimension with superposed segments on multiple layers that are stapled together with vias. In addition, the width determination is typically an iterative process, since currents depend on timing and noise, which depend on voltage drops, which, in turn, depend on currents. In other words, there is a cyclic dependency within the power, timing and noise analyses. This loop is typically addressed through multiple iterations and by experienced designers. After executing the above three steps, the segments of the power-ground route form obstacles during general signal routing (Fig. 3.22(b)).

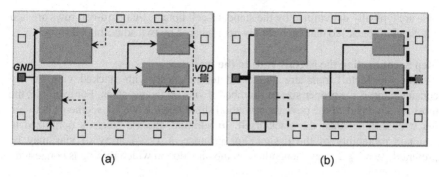

Fig. 3.22 (a) Generating the topology of the two supply nets. (b) Adjusting the widths of the individual segments in (a) with respect to their maximum current loads.

► 3.7.3 Mesh Routing

Power-ground routing in modern digital ICs typically has a *mesh* topology that is created through the following five steps.

Step 1: creating a ring
Typically, a ring is constructed to surround the entire core area of the chip, and possibly individual blocks. The purpose of the ring is to connect the supply I/O cells and, possibly, electrostatic discharge protection structures with the global power mesh of the chip or block. For low resistance, these connections and the ring itself are on many layers. For example, a ring might use metal layers Metal2- Metal8 (every layer except Metal1).

Step 2: connecting I/O pads to the ring
The top left of Fig. 3.23 shows connectors from the I/O pads to the ring. Each I/O pad will have a number of "fingers" emanating on each of several metal layers. These should be maximally connected to the power ring in order to minimize resistance and maximize the ability to carry current to the core.

Step 3: creating a mesh
A power mesh consists of a set of stripes at defined pitches on two or more layers (Fig. 3.23). The width and pitch of stripes are determined from estimated power consumption as well as layout design rules. The stripes are laid out in pairs, alternating as *VDD-GND*, *VDD-GND*, and so on. The power mesh uses the uppermost and thickest layers, and is sparser on any lower layers to avoid signal routing congestion. Stripes on adjacent layers are typically connected with as many vias as possible, again to minimize resistance.

Step 4: creating Metal1 rails
The Metal1 layer is where the power-ground distribution network meets the logic gates of the design. The width (current supply capability) and pitch of the Metal1

rails are typically determined by the standard-cell library. Standard-cell rows are laid out "back-to-back" so that each supply net is shared by two adjacent cell rows.

Step 5: connecting the Metal1 rails to the mesh

Finally, the Metal1 rails are connected to the mesh with stacked vias. A key consideration is the proper size of (number of vias in) the via stack. For example, the most resistive part of the power distribution should be the Metal1 segments between via stacks, rather than the stack itself. In addition, the via stack is optimized to maintain routability of the design. For instance, a 1×4 array of vias may be preferable to a 2×2 array, depending on the direction in which routing is congested.

Figure 3.23 illustrates the mesh approach to power-ground distribution. In the figure, layers Metal8 through Metal4 are used for the mesh. In practice, many chips will use fewer layers (e.g., Metal8 and Metal7 only) due to routing resource constraints.

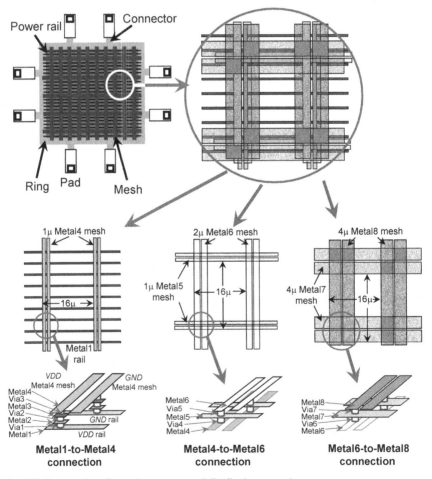

Fig. 3.23 Construction of a mesh power-ground distribution network.

Chapter 3 Exercises

Exercise 1: Slicing Trees and Constraint Graphs
For the given floorplan (right), generate its slicing tree, vertical constraint graph and horizontal constraint graph.

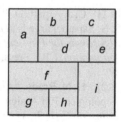

Exercise 2: Floorplan-Sizing Algorithm
Three blocks *a*, *b* and *c* are given along with their size options.

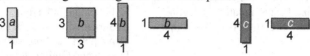

(a) Determine the shape functions for each block *a, b, c*.
(b) Find the minimum area of the top-level floorplan using the given tree structure and determine the shape function of the top-level floorplan. In the shape function, find the corner point that yields the minimal area. Finally, determine the dimensions of each block and draw the resulting floorplan.

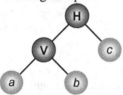

Exercise 3: Linear-Ordering Algorithm
For the netlist with five blocks *a-e* and six nets N_1-N_6, determine the linear ordering that minimizes total wirelength. Let the starting block be block *a*. Place it in the first (leftmost) position. Draw the resulting placement.

$N_1 = (a,e)$ $N_2 = (a,b)$
$N_3 = (a,c,d)$ $N_4 = (a,d)$
$N_5 = (b,c,d)$ $N_6 = (b,d)$

Exercise 4: Non-Slicing Floorplans
Recall that the smallest non-slicing floorplans with *no* wasted space exhibit the structure of a clockwise or counter-clockwise *wheel* with five blocks. Draw a non-slicing floorplan with only four blocks *a-d*.

Chapter 3 References

[3.1]   C. J. Alpert, D. P. Mehta and S. S. Sapatnekar, eds., *Handbook of Algorithms for Physical Design Automation*, CRC Press, 2009.

[3.2]   H. N. Brady, "An Approach to Topological Pin Assignment", *IEEE Trans. on CAD* 3(3) (1984), pp. 250-255.

[3.3]   T.-C. Chen and Y.-W. Chang, "Modern Floorplanning Based on B*-Tree and Fast Simulated Annealing", *IEEE Trans. on CAD* 25(4) (2006), pp. 637-650.

[3.4]   S. Kang, "Linear Ordering and Application to Placement", *Proc. Design Autom. Conf.*, 1983, pp. 457-464.

[3.5]   S. Kirkpatrick, C. D. Gelatt and M. P. Vecchi, "Optimization by Simulated Annealing", *Science* 220(4598) (1983), pp. 671-680.

[3.6]   N. L. Koren, "Pin Assignment in Automated Printed Circuit Board Design", *Proc. Design Autom. Workshop*, 1972, pp. 72-79.

[3.7]   M. D. Moffitt, J. A. Roy, I. L. Markov and M. E. Pollack, "Constraint-Driven Floorplan Repair", *ACM Trans. on Design Autom. of Electronic Sys.* 13(4) (2008), pp. 1-13.

[3.8]   R. H. J. M. Otten, "Efficient Floorplan Optimization", *Proc. Intl. Conf. on Computer Design*, 1983, pp. 499-502.

[3.9]   R. H. J. M. Otten and L. P. P. P. van Ginneken, "Floorplan Design Using Simulated Annealing", *Proc. Intl. Conf. on CAD*, 1984, pp. 96-98.

[3.10] C. Sechen, "Chip Planning, Placement and Global Routing of Macro/Custom Cell Integrated Circuits Using Simulated Annealing", *Proc. Design Autom. Conf.*, 1988, pp. 73-80.

[3.11] L. Stockmeyer, "Optimal Orientation of Cells in Slicing Floorplan Designs", *Information and Control* 57 (1983), pp. 91-101.

[3.12] S. Sutanthavibul, E. Shragowitz and J. B. Rosen, "An Analytical Approach to Floorplan Design and Optimization", *IEEE Trans. on CAD* 10 (1991), pp. 761-769.

[3.13] X. Tang, R. Tian and D. F. Wong, "Fast Evaluation of Sequence Pair in Block Placement by Longest Common Subsequence Computation", *Proc. Design, Autom. and Test in Europe*, 2000, pp. 106-111.

[3.14] D. F. Wong and C. L. Liu, "A New Algorithm for Floorplan Design", *Proc. Design Autom. Conf.*, 1986, pp. 101-107.

[3.15] J. Xiong, Y.-C. Wong, E. Sarto and L. He, "Constraint Driven I/O Planning and Placement for Chip-Package Co-Design", *Proc. Asia and South Pacific Design Autom. Conf.*, 2006, pp. 207-212.

[3.16] J. Z. Yan and C. Chu, "DeFer: Deferred Decision Making Enabled Fixed-Outline Floorplanner", *Proc. Design Autom. Conf.*, 2008, pp. 161-166.

[3.17] T. Yan and H. Murata, "Fast Wire Length Estimation by Net Bundling for Block Placement", *Proc. Intl. Conf. on CAD*, 2006, pp. 172-178.

Chapter 4
Global and Detailed Placement

4

4

4 Global and Detailed Placement

After partitioning the circuit into smaller modules and floorplanning the layout to determine block outlines and pin locations, *placement* seeks to determine the locations of standard cells or logic elements within each block while addressing optimization objectives, e.g., minimizing the total length of connections between elements. Specifically, *global placement* (Sec. 4.3) assigns general locations to movable objects, while *detailed placement* (Sec. 4.4) refines object locations to legal cell sites and enforces nonoverlapping constraints. The detailed locations enable more accurate estimates of circuit delay for the purpose of timing optimization.

4.1 Introduction

The objective of placement is to determine the locations and orientations of all circuit elements within a (planar) layout, given solution constraints (e.g., no overlapping cells) and optimization goals (e.g., minimizing total wirelength).

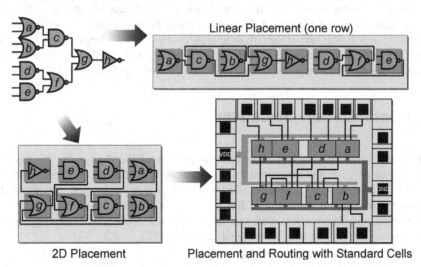

Fig. 4.1 A simple circuit (top left) along with example linear placement (top right), 2D placement (bottom left), and placement and routing with standard cells (bottom right).

Circuit elements (e.g., gates, standard cells and macro blocks) have rectangular shapes and are represented by nodes, while nets are represented by edges (Fig. 4.1). Some circuit elements may have fixed locations while others are movable (placeable). The placement of movable objects will determine the achievable quality of the subsequent routing stages. However, detailed routing information, such as track assignment, is not available at the placement stage, and hence the placer estimates the eventual *Manhattan* routing, where only horizontal and vertical wires are allowed (Sec. 4.2). In Fig. 4.1, assuming unit distance between horizontally- or vertically-adjacent placement sites, both the linear and 2D placements have 10 units of total wirelength.

Placement techniques for large circuits encompass *global placement, detailed placement* and legalization. Global placement often neglects specific shapes and sizes of placeable objects and does not attempt to align their locations with valid grid rows and columns. Some overlaps are allowed between placed objects, as the emphasis is on judicious global positioning and overall density distribution. Legalization is performed before or during detailed placement. It seeks to align placeable objects with rows and columns, and remove overlap, while trying to minimize displacements from global placement locations as well as impacts on interconnect length and circuit delay. Detailed placement incrementally improves the location of each standard cell by local operations (e.g., swapping two objects) or shifting several objects in a row to create room for another object. Global and detailed placement typically have comparable runtimes, but global placement often requires much more memory and is more difficult to parallelize.

Performance-driven optimizations can be applied during both global placement and detailed placement. However, timing estimation (Chap. 8) can be inaccurate during early stages of global placement. Thus, it is more common to initiate performance optimizations during later stages of, or after, global placement. Legalization is often performed so as to minimize impact on performance, and detailed placement can directly improve performance because of its fine control over individual locations.

4.2 Optimization Objectives

Placement must produce a layout wherein all nets of the design can be routed simultaneously, i.e., the placement must be *routable*. In addition, electrical effects such as signal delay or crosstalk must be taken into consideration. As detailed routing information is not available during placement, the placer optimizes estimates of routing quality metrics, such as total weighted wirelength, cut size, wire congestion (density), or maximum signal delay (Fig. 4.2).

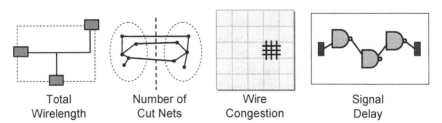

| Total
Wirelength | Number of
Cut Nets | Wire
Congestion | Signal
Delay |

Fig. 4.2 Examples of routing quality metrics optimized during placement.

Subject to maintaining routability, the primary goal of placement is to optimize delays along signal paths. Since the delay of a net directly correlates with the net's length, placers often minimize *total wirelength*.[1] As illustrated in Fig. 4.3, the placement of a design strongly affects the net lengths as well as the wire density. Further placement optimizations, beyond the scope of this chapter, include

– Placement and assignment of I/O pads with respect to both the logic gates connected to them and the package connection points, e.g., bump locations.
– Temperature- and reliability-driven optimizations, such as the placement of actively switching (and heat-generating) circuit elements, to achieve uniform temperature across the chip.

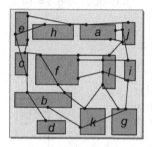 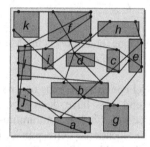

Fig. 4.3 Two placements of the same design with better (left) and worse (right) total wirelength.

Wirelength estimation for a given placement. An important consideration for wirelength estimation during placement is the speed at which a given wirelength estimate can be computed. Moreover, the estimation must be applicable to both *two-pin* and *multi-pin nets*. For two-pin nets, most placement tools use the Manhattan distance d_M between two points (pins) P_1 (x_1, y_1) and P_2 (x_2, y_2)

$$d_M(P_1, P_2) = |x_2 - x_1| + |y_2 - y_1|$$

For multi-pin nets, the following estimation techniques are used [4.23].

The *half-perimeter wirelength* (*HPWL*) model is commonly used because it is reasonably accurate and efficiently calculated. The *bounding box* of a net with p pins is the smallest rectangle that encloses the pin locations. The wirelength is estimated as half the perimeter of the bounding box. For two- and three-pin nets (70-80% of all nets in most modern designs), this is exactly the same as the rectilinear Steiner minimum tree (RSMT) cost (discussed later in this section). When $p \geq 4$, HPWL underestimates the RSMT cost by an average factor that grows asymptotically as $\sqrt{p}$.

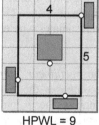

HPWL = 9

---

[1] Timing-driven placement techniques are discussed in Sec. 8.3.

The *complete graph* (*clique*) model of a *p*-pin net *net* has

$$\binom{p}{2} = \frac{p!}{2!(p-2)!} = \frac{p(p-1)}{2}$$

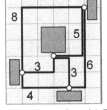

edges, i.e., each pin is directly connected to every other pin. Since a spanning tree over the net's pins will have $p - 1$ edges, a correction factor of $2 / p$ is applied. The total edge length of *net* according to the clique model is

$$L(net) = \frac{2}{p} \sum_{e \in clique} d_M(e)$$

Clique Length = 14.5

where e is an edge in the clique, and $d_M(e)$ is the Manhattan distance between the endpoints of e.

The *monotone chain* model connects the pins of a net using a chain topology. Both end pins have degree one; each intermediate pin has degree two. Finding a minimum-length path connecting the pin locations corresponds to the NP-hard Hamiltonian path problem. Thus, the monotone chain model sorts pins by either *x*- or *y*-coordinate and connects them accordingly. Though simple, this method often overestimates the actual wirelength. Another disadvantage is that the chain topology changes with the placement.

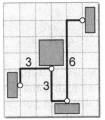

Chain Length = 12

The *star* model considers one pin as the *source* node and all other pins as *sink* nodes; there is an edge from the source to each sink. This is especially useful for timing optimization, since it captures the direction of signal flow from an output pin to one or more input pins. The star model uses only $p - 1$ edges; this sparsity can be advantageous in modeling high pin-count nets. On the other hand, the star model overestimates wirelength.

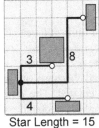

Star Length = 15

The *rectilinear minimum spanning tree* (*RMST*) model decomposes the *p*-pin net into two-pin connections and connects the *p* pins with $p - 1$ connections. Several algorithms (e.g., *Kruskal's Algorithm* [4.19]) exist for constructing minimum spanning trees. RMST algorithms can exploit the Manhattan geometry to achieve $O(p \log p)$ runtime complexity (Sec. 5.6.1).

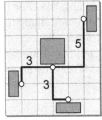

RMST Length = 11

The *rectilinear Steiner minimum tree* (*RSMT*) model connects all p pins of the net and as many as $p - 2$ additional *Steiner* (branching) *points* (Sec. 5.6.1). Finding an optimal set of Steiner points for an arbitrary point set is NP-hard. For nets with a bounded number of pins, computing an RSMT takes constant time. If the Steiner points are known, an RSMT can be found by constructing an RMST over the union of the original point set and the set of added Steiner points.

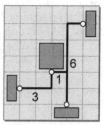

RSMT Length = 10

The *rectilinear Steiner arborescence* (*RSA*) model of a p-pin net is also a tree where a single source node s_0 is connected to $p - 1$ sink nodes. In an RSA, the path length from s_0 to any sink s_i, $1 \le i < p - 1$, must be equal to the $s_0 \sim s_i$ Manhattan (rectilinear) distance. That is, for all sinks s_i

$$L(s_0, s_i) = d_M(s_0, s_i)$$

where $L(s_0, s_i)$ is the path length from s_0 to s_i in the tree. Computing a minimum-length RSA is NP-hard.

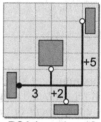

RSA Length = 10

The *single-trunk Steiner tree* (*STST*) model consists of one vertical (horizontal) segment, i.e., *trunk*, and connects all pins to this trunk using horizontal (vertical) segments, i.e., *branches*. STSTs are commonly used for estimation due to their ease of construction. RSAs are more timing-relevant but somewhat more complex to construct than STSTs. For practical purposes, both RSAs and STSTs are constructed in $O(p \log p)$-time, where p is the number of pins.

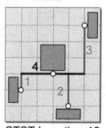

STST Length = 10

Total wirelength with net weights (weighted wirelength). *Net weights* can be used to prioritize certain nets over others. For example, a net *net* with weight $w(net) = 2$ is equivalent to two nets i and j with weights $w(i) = w(j) = 1$. For a placement P, an estimate of total weighted wirelength is

$$L(P) = \sum_{net \in P} w(net) \cdot L(net)$$

where $w(net)$ is the weight of *net*, and $L(net)$ is the estimated wirelength of *net*.

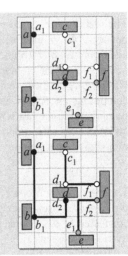

Example: Total Weighted Wirelength of a Placement
Given: (1) placement P of blocks a-f and their pins (right) and
(2) nets N_1-N_3 and their net weights.

$$N_1 = (a_1,b_1,d_2) \qquad w(N_1) = 2$$
$$N_2 = (c_1,d_1,f_1) \qquad w(N_2) = 4$$
$$N_3 = (e_1,f_2) \qquad w(N_3) = 1$$

Task: estimate the total weighted wirelength of P using the RMST model.

Solution:
$$L(N_1) = d_M(a_1,b_1) + d_M(b_1,d_2) = 4 + 3 = 7.$$
$$L(N_2) = d_M(c_1,d_1) + d_M(d_1,f_1) = 2 + 2 = 4.$$
$$L(N_3) = d_M(e_1,f_2) = 3.$$

$$L(P) = w(N_1) \cdot L(N_1) + w(N_2) \cdot L(N_2) + w(N_3) \cdot L(N_3)$$
$$= 2 \cdot 7 + 4 \cdot 4 + 1 \cdot 3 = 14 + 16 + 3 = 33.$$

Maximum cut size. In Fig. 4.4, a (global) vertical cutline divides the region into a left region L and a right region R. With respect to the cutline, a net can be classified as either *uncut* or *cut*. An uncut net has pins in L or R, but not both, i.e., it is completely left or right of the cutline; a cut net has at least one pin in each L and R.

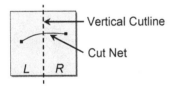

Fig. 4.4 Layout with a vertical cutline and a cut net.

Given a placement P, let

- V_P and H_P be the set of global vertical and horizontal cutlines of P, respectively
- $\Psi_P(cut)$ be the set of nets cut by a cutline *cut*
- $\psi_P(cut)$ be the size of $\Psi_P(cut)$, i.e., $\psi_P(cut) = |\Psi_P(cut)|$

Then, define $X(P)$ to be the maximum of $\psi_P(v)$ over all vertical cutlines $v \in V_P$.

$$X(P) = \max_{v \in V_P}\big(\psi_P(v)\big)$$

Similarly, define $Y(P)$ as the maximum of $\psi_P(h)$ over all horizontal cutlines $h \in H_P$.

$$Y(P) = \max_{h \in H_P}\big(\psi_P(h)\big)$$

$X(P)$ and $Y(P)$ are lower bounds on routing capacity needed in the horizontal (x-) and vertical (y-) directions, respectively. For example, if $X(P) = 10$, then some global vertical cutline x crosses 10 horizontal net segments. Likewise, if $Y(P) = 15$, some global horizontal cutline y crosses 15 vertical net segments. A necessary but not sufficient condition for routability is that there exist at least 10 (15) horizontal (vertical) routing tracks at x (y). Thus, $X(P)$ and $Y(P)$ can be used to assess routability of P.

For some circuit layout styles, e.g., gate arrays, the *capacity* (maximum number) of horizontal and vertical tracks is pre-set. An optimization constraint in placement, necessary but not sufficient for routability, is to ensure that $X(P)$ and $Y(P)$ are within these capacities. In the context of standard-cell design, $X(P)$ gives a lower bound on the demand for horizontal routing tracks, and $Y(P)$ gives a lower bound on the demand for vertical routing tracks.

To improve routability of P, $X(P)$ and $Y(P)$ must be minimized. To improve total wirelength of P, separately calculate the number of crossings of global vertical and horizontal cutlines, and minimize

$$L(P) = \sum_{v \in V_P} \psi_P(v) + \sum_{h \in H_P} \psi_P(h)$$

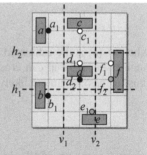

Example: Cut Sizes of a Placement
Given: (1) placement P of blocks a-f and their pins (right), (2) nets N_1-N_3, (3) global vertical cutlines v_1 and v_2, and (4) global horizontal cutlines h_1 and h_2.
$\quad N_1 = (a_1, b_1, d_2) \quad N_2 = (c_1, d_1, f_1) \quad N_3 = (e_1, f_2)$
Task: determine the cut sizes $X(P)$ and $Y(P)$ of placement P according to the RMST model.

Solution:
Find the cut values for each global cutline.
$\quad \psi_P(v_1) = 1 \quad \psi_P(v_2) = 2$
$\quad \psi_P(h_1) = 3 \quad \psi_P(h_2) = 2$

Find the total number of crossings in P.
$\psi_P(v_1) + \psi_P(v_2) + \psi_P(h_1) + \psi_P(h_2) = 1 + 2 + 3 + 2 = 8$

Find the cut sizes.
$X(P) = \max(\psi_P(v_1), \psi_P(v_2)) = \max(1,2) = 2$
$Y(P) = \max(\psi_P(h_1), \psi_P(h_2)) = \max(3,2) = 3$

Observe that moving block b from $(0,0)$ to $(0,1)$ reduces the number of crossings from 8 to 6. This also decreases $\psi_P(h_1)$ from 3 to 1, which makes $Y(P) = 2$, thereby reducing the local congestion.

Routing congestion. The routing congestion of a placement P can be thought of in terms of *density*, namely, the ratio of *demand* for routing tracks to the *supply* of available routing tracks. For example, a *routing channel* has available horizontal routing tracks, while a *switchbox* has available vertical and horizontal routing tracks

(Fig. 4.5). For gate-array designs, this concept of congestion is particularly important, as the supply of routing tracks is fixed (Sec. 5.3).

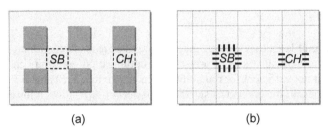

(a)   (b)

Fig. 4.5 (a) A switchbox *SB* and a channel *CH*. (b) *SB* and *CH* represented by grid cells with wire capacities on the grid cell edges.

For a given placement P, congestion can also be estimated by the number of nets that pass through the boundaries of individual routing regions. Formally, the *local wire density* $\varphi_P(e)$ of an edge e between two neighboring grid cells[2] is

$$\varphi_P(e) = \frac{\eta_P(e)}{\sigma_P(e)}$$

where $\eta_P(e)$ is the estimated number of nets that cross e and $\sigma_P(e)$ is the maximum number of nets that can cross e. If $\varphi_P(e) > 1$, then too many nets are estimated to cross e, making P more likely to be unroutable. The *wire density* of P is

$$\Phi(P) = \max_{e \in E}(\varphi_P(e))$$

where E is the set of all edges. If $\Phi(P) \leq 1$, then the design is estimated to be fully routable. If $\Phi(P) > 1$, then routing will need to detour some nets through less-congested edges – but in some cases, this may be impossible. Therefore, congestion-driven placement often seeks to minimize $\Phi(P)$.

Example: Wire Density of a Placement
Given: (1) placement P of blocks *a-f* and their pins (right), (2) nets N_1-N_3, (3) local vertical cutlines v_1-v_6, (4) local horizontal cutlines h_1-h_6, and (5) $\sigma_P(e) = 3$ for all local cutlines $e \in E$.
$N_1 = (a_1, b_1, d_2)$ $N_2 = (c_1, d_1, f_1)$ $N_3 = (e_1, f_2)$
Task: find the wire density $\Phi(P)$ and determine the routability of P based on the RMST model.

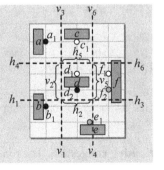

---

[2]   Such an edge represents the border between two regions (e.g., switchboxes or channels) or between two cutlines.

Solution:
The following is one of many possible solutions, as $\Phi(P)$ depends on how N_1-N_3 are routed.

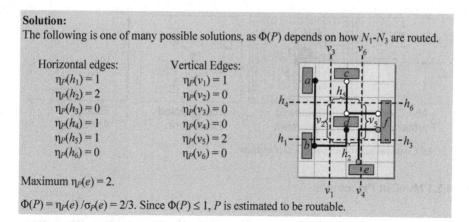

Horizontal edges:	Vertical Edges:
$\eta_P(h_1) = 1$	$\eta_P(v_1) = 1$
$\eta_P(h_2) = 2$	$\eta_P(v_2) = 0$
$\eta_P(h_3) = 0$	$\eta_P(v_3) = 0$
$\eta_P(h_4) = 1$	$\eta_P(v_4) = 0$
$\eta_P(h_5) = 1$	$\eta_P(v_5) = 2$
$\eta_P(h_6) = 0$	$\eta_P(v_6) = 0$

Maximum $\eta_P(e) = 2$.

$\Phi(P) = \eta_P(e) / \sigma_P(e) = 2/3$. Since $\Phi(P) \leq 1$, P is estimated to be routable.

Signal delays. The total wirelength of the placement affects the maximum clock frequency for a given design, as it depends on net (wire) delays and gate delays. In earlier process technologies, gate delays accounted for the majority of circuit delay. With modern processes, however, due to technology scaling, wire delays contribute a significant portion of overall path delay.

Circuit timing is usually verified using *static timing analysis* (*STA*) (Sec. 8.2.1), based on estimated net and gate delays. Common terminology includes *actual arrival time* (*AAT*) and *required arrival time* (*RAT*), which can be estimated for every node v in the circuit. $AAT(v)$ represents the latest transition time at a given node v measured from the beginning of the clock cycle. $RAT(v)$ represents the time by which the latest transition at v must complete in order for the circuit to operate correctly within a given clock cycle. For correct operation of the chip with respect to setup (maximum path delay) constraints, it is required that $AAT(v) \leq RAT(v)$.

4.3 Global Placement

4.3

Techniques for circuit placement are summarized as follows (Fig. 4.6). In *partitioning-based algorithms*, the netlist and the layout are divided into smaller sub-netlists and sub-regions, respectively, according to cut-based cost functions. This process is repeated until each sub-netlist and sub-region is small enough to be handled optimally. An example of this approach is *min-cut placement* (Sec. 4.3.1).

Analytic techniques model the placement problem using an objective (cost) function, which can be maximized or minimized via mathematical analysis. The objective can be quadratic or otherwise non-convex. Examples of analytic techniques include *quadratic placement* and *force-directed placement* (Sec. 4.3.2).

In *stochastic algorithms*, randomized moves are used to optimize the cost function. An example of this approach is *simulated annealing* (Sec. 4.3.3).

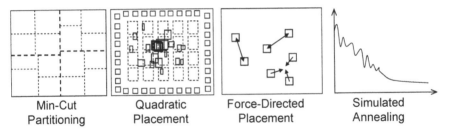

| Min-Cut
Partitioning | Quadratic
Placement | Force-Directed
Placement | Simulated
Annealing |

Fig. 4.6 Common techniques for global placement.

▶ 4.3.1 Min-Cut Placement

In the late 1970s, *M. A. Breuer* [4.1] studied *min-cut placement*, which uses partitioning algorithms to divide (1) the netlist and (2) the layout region into smaller, sub-netlists and sub-regions, respectively. The sub-netlists and sub-regions are repeatedly divided into even smaller partitions until each sub-region contains only a small number of cells. Conceptually, each sub-region is assigned a portion of the original netlist. However, when implementing min-cut placement, the netlist should be divided such that each sub-region has access to its own unique (induced) sub-netlist.

Each cut heuristically minimizes the number of cut nets (Sec. 4.2). Standard algorithms used to minimize the number of cut nets are the Kernighan-Lin (KL) algorithm (Sec. 2.4.1) and the Fiduccia-Mattheyses (FM) algorithm (Sec. 2.4.3).

Min-Cut Algorithm
Input: netlist *Netlist*, layout area *LA*, minimum number of cells per region *cells_min*
Output: placement *P*
1. $P = \varnothing$
2. *regions* = ASSIGN(*Netlist*,*LA*)   // assign netlist to layout area
3. **while** (*regions* != $\varnothing$)   // while regions still not placed
4.   *region* = FIRST_ELEMENT(*regions*)   // first element in *regions*
5.   REMOVE(*regions*, *region*)   // remove first element of *regions*
6.   **if** (*region* contains more than *cell_min* cells)
7.   (*sr₁*,*sr₂*) = BISECT(*region*)   // divide *region* into two subregions
                                                         //   *sr₁* and *sr₂*, obtaining the sub-
                                                         //   netlists and sub-areas
8.   ADD_TO_END(*regions*,*sr₁*)   // add *sr₁* to the end of *regions*
9.   ADD_TO_END(*regions*,*sr₂*)   // add *sr₂* to the end of *regions*
10.   **else**
11.   PLACE(*region*)   // place *region*
12.   ADD(*P*,*region*)   // add *region* to *P*

Min-cut optimization is performed *iteratively*, one cutline at a time. That is, heuristic minimum cuts are found for the current sub-netlist, based on the current sub-regions. Ideally, the algorithm should directly optimize the placement figures of merit $X(P)$,

$Y(P)$ and $L(P)$ (Sec. 4.2). However, this is computationally infeasible, especially for large designs. Moreover, even if every cut were optimal, the final solution would not necessarily be optimal. Therefore, the algorithm iteratively minimizes cut size, where the cut size is minimized on the first (horizontal) cut cut_1, then minimized on the second (vertical) cut cut_2, and so on.

$$\text{minimize}(\psi_P(cut_1)) \rightarrow \text{minimize}(\psi_P(cut_2)) \rightarrow \ldots \rightarrow \text{minimize}(\psi_P(cut_{|Cuts|}))$$

Let *Cuts* be the set of all cutlines made in the layout region. The sequence cut_1, cut_2, ... , $cut_{|Cuts|}$ denotes the order in which the cuts are made. Possible approaches to dividing the layout include using alternating and repeating cutline directions.

With *alternating cutline directions*, the algorithm divides the layout by switching between sets of vertical and horizontal cutlines. In Fig. 4.7(a), the horizontal cutline cut_1 is made first to bisect the region. Then, two vertical cutlines cut_{2a} and cut_{2b} are made, one (cut_{2a}) in the top half and the other (cut_{2b}) in the bottom half. Each newly formed region is divided in half again with four horizontal cutlines cut_{3a}-cut_{3d}, followed by eight vertical cutlines cut_{4a}-cut_{4h}. This approach is suitable for standard-cell designs with high wire density in the center of the layout region.

With *repeating cutline directions*, the layout is divided using only vertical (horizontal) cutlines until each column (row) is the width (height) of a standard cell. Then, the layout is divided using the orthogonal set of cutlines, e.g., if horizontal cutlines are first used to generate rows, then vertical lines are used to divide each row into columns. In Fig. 4.7(b), the horizontal cut cut_1 is made first to divide the region in half. Then, cut_{2a} and cut_{2b} divided the region into four rows. Next, four vertical cuts cut_{3a}- cut_{3d} are made, followed by eight vertical cuts cut_{4a}- cut_{4h}. This approach often results in greater wirelength because the aspect ratios of the sub-regions can be very far from one. For example, when bisecting the sub-netlist with cut_{3a}, very little information is available regarding the x-locations of adjacent standard cells in the other sub-regions.

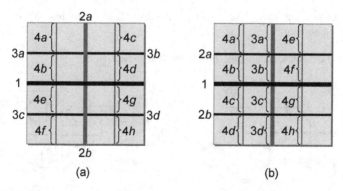

(a)   (b)

Fig. 4.7 Partitioning a region using different cutline approaches. (a) Alternating cutline directions. (b) Repeating cutline directions.

Example: Min-Cut Placement Using the KL Algorithm

Given: (1) circuit with gates *a-f* (left), (2) 2 × 4 layout (right), and (3) initial vertical cut cut_1.

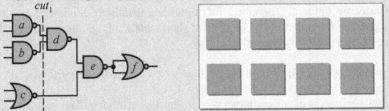

Task: find a placement with minimum wirelength using alternating cutline directions and the KL algorithm.

Solution:
After vertical cut cut_1, $L = \{a,b,c\}$ and $R = \{d,e,f\}$. Partition using the KL algorithm.

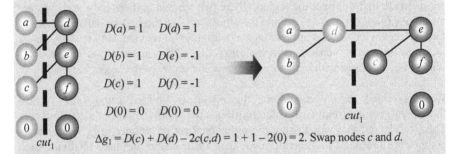

$D(a) = 1$ $D(d) = 1$

$D(b) = 1$ $D(e) = -1$

$D(c) = 1$ $D(f) = -1$

$D(0) = 0$ $D(0) = 0$

$\Delta g_1 = D(c) + D(d) - 2c(c,d) = 1 + 1 - 2(0) = 2$. Swap nodes *c* and *d*.

After horizontal cut cut_{2L}, $T = \{a,d\}$, $B = \{b\}$.   After horizontal cut cut_{2R}, $T = \{c,e\}$, $B = \{f\}$.

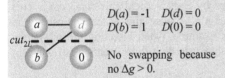

$D(a) = -1$ $D(d) = 0$
$D(b) = 1$ $D(0) = 0$

No swapping because no $\Delta g > 0$.

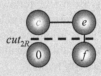

$D(c) = -1$ $D(e) = 0$
$D(0) = 0$ $D(f) = 1$

No swapping because no $\Delta g > 0$.

Make four vertical cuts cut_{3TL}, cut_{3TR}, cut_{3BL} and cut_{3BR}. Each region has only one node, so terminate the algorithm.

After external pin consideration (not shown), the final placement is

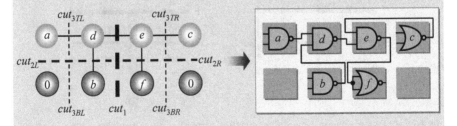

Example: Min-Cut Placement Using the FM Algorithm
Given: (1) circuit with gates *a-g* (left), (2) gate areas, (3) ratio factor $r = 0.5$, (4) initial partitioning with vertical cut cut_1, and (5) a 2×4 layout (right).

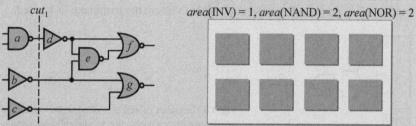

$area(\text{INV}) = 1, area(\text{NAND}) = 2, area(\text{NOR}) = 2$

Task: find a placement with minimum wirelength using alternating cutline directions and using the FM algorithm.

Solution:
Initial vertical cut cut_1: $L = \{a,b,c\}$, $R = \{d,e,f,g\}$.
Balance criterion: $0.5 \cdot 11 - 2 \leq area(A) \leq 0.5 \cdot 11 + 2 = 3.5 \leq area(A) \leq 7.5$.

Iteration 1:
Gates *a*, *b*, *c* and *g* have maximum gain $\Delta g_1 = 1$.
Balance criterion for gates *a*, *b* and *c* is violated: $area(A) < 3.5$.
Balance criterion for gate *g* is met: $area(A) = 6$.
Move gate *g*.

Iteration 2:
Gate *a* has maximum gain $\Delta g_2(a) = 1$, $area(A) = 4$, balance criterion is met.
Move gate *a* (further selection steps result in negative Δg and are omitted).
Maximum positive gain $G_2 = \Delta g_1 + \Delta g_2 = 2$.
After cut cut_1, $L = \{a,d,e,f\}$, $R = \{b,c,g\}$, cut cost = 1.

After cut cut_{2T}, $T = \{a,d\}$, $B = \{e,f\}$, cut cost = 1.

After cut cut_{2B}, $T = \{c\}$, $B = \{b,g\}$, cut cost = 1.

Make three more cuts cut_{3TL}, cut_{3TR} and cut_{3BR} such that every sub-region has one gate.

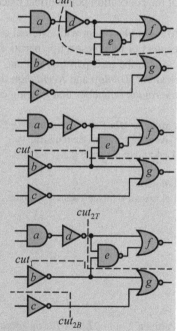

After external pin consideration (not shown), the final placement is

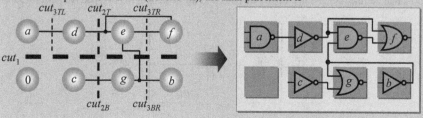

Naive min-cut algorithms do not consider the locations of connection pins within partitions that have already been visited. Likewise, the locations of fixed, external pin connections (pads) are also ignored. However, as illustrated in Fig. 4.8, cell *a* should properly be placed as close as possible to where the terminal *p'* is located.

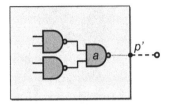

Fig. 4.8 Placement of cell *a* is close to the terminal *p'*, which represents a connection to a neighboring partition.

Formally developed by *A. E. Dunlop* and *B. W. Kernighan* [4.8], *terminal propagation* considers external pin locations during partitioning-based placement. During min-cut placement, external connections are represented by artificial connection points on the cutline and are dummy nodes in hypergraphs. The locations of the connection points affect placement cost functions and hence cell placements.

Min-cut placement with external connections assumes that the cells are placed in the centers of their respective partitions. If the related connections (dummy nodes) are *close* to the next partition cutline, these nodes are not considered when making the next cut. *Dunlop* and *Kernighan* define close as *projecting onto the middle third of the region boundary* that is being cut [4.8].

Example: Min-Cut Placement With External Connections
Given: gates *a-d* of a circuit (right).
Task: place the gates in a 2 × 2 grid.

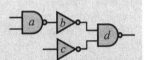

Solution:
Let gates *a-d* be represented by nodes *a-d*. Partition into *L* and *R*. Partition cost = 1.

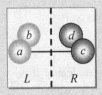

Partition *L* into *TL* and *BL independent* of *R*, since the connection *x* is close to the new horizontal cutline (assume that nodes are placed in the centers of their respective partitions).

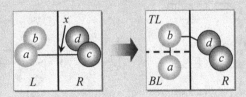

Partition R into TR and BR *depending* on L, since the connection x is not close to the cutline. The addition of the dummy node p' means that node d moves to grid TR and node c moves to grid BR so that the partition cost of the TR-BR partition cost = 1.

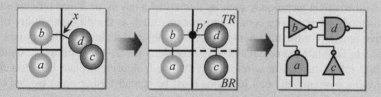

Note that without p', the cost of placing c in TR and node d in BR is still 1, but the total cost with consideration of L becomes 2.

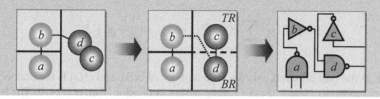

If a net *net* crosses the partitioned region, all its pins, including those that lie outside the region, must be considered. A rectilinear Steiner minimum tree (RSMT) (Sec. 4.2) is constructed that includes each pin outside the region. Each point at which this tree crosses the region's boundary induces a dummy node p' for *net*. If p' is close to a cutline, then it is ignored during partitioning. Otherwise, it is considered during partitioning along with cells that are contained in the region.

Example: Min-Cut Placement Considering Pins Outside of the Partitioned Region
Given: (1) layout region (right) and (2) net N_1 with three pins a-c not in the partitioned region.
Task: account for pins a-c.

Solution:

For pins a-c, construct a rectilinear Steiner minimum tree (RSMT). Label all points of that tree that cross the layout region with dummy nodes p_a'-p_c'.

If a vertical cut cut_V is made, then p_b' is ignored because it is *close* to the cutline, while p_a' and p_c' are considered as real nodes. If a horizontal cut cut_H is made, then p_c' is ignored, and p_a' and p_b' are considered as real nodes.

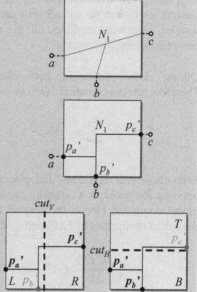

▶ **4.3.2 Analytic Placement**

Analytic placement minimizes a given objective, such as wirelength or circuit delay, using mathematical techniques such as numerical analysis or linear programming. Such methods often require certain assumptions, such as the differentiability of the objective function or the treatment of placeable objects as dimensionless points. For example, to facilitate the calculation of partial derivatives, it is common to optimize *quadratic*, rather than linear wirelength. When such algorithms place cells too close, i.e., creating overlaps, the cell locations must be spread further apart by dedicated post-processing techniques, so as to remove overlap.

Quadratic placement. The *squared* Euclidean distance

$$L(P) = \sum_{i=1,j=1}^{n} c(i, j)\left((x_i - x_j)^2 + (y_i - y_j)^2\right)$$

is used as the cost function, n is the total number of cells, and $c(i,j)$ is the connection cost between cells i and j. If cells i and j are not connected, then $c(i,j) = 0$. The terms $(x_i - x_j)^2$ and $(y_i - y_j)^2$ respectively give the squared horizontal and vertical distances between the centers of i and j. This formulation implicitly decomposes all nets into two-pin subnets. The quadratic form emphasizes the minimization of long connections, which tend to have negative impacts on timing.

Quadratic placement consists of two stages. During *global placement* (first stage), cells are placed so as to minimize the quadratic function with respect to the cell centers. Note that this placement is not legal. Usually, cells appear in large clusters with many cell overlaps. During *detailed placement* (second stage), these large clusters are broken up and all cells are placed such that no overlap occurs. That is, detailed placement legalizes all the cell locations and produces a high-quality, non-overlapping placement.

During *global placement*, each dimension can be considered independently. Therefore, the cost function $L(P)$ can be separated into x- and y-components

$$L_x(P) = \sum_{i=1,j=1}^{n} c(i, j)(x_i - x_j)^2 \quad \text{and} \quad L_y(P) = \sum_{i=1,j=1}^{n} c(i, j)(y_i - y_j)^2$$

With these cost functions, the placement problem becomes a convex quadratic optimization problem. Convexity implies that any local minimum solution is also a global minimum. Hence, the optimal x- and y-coordinates can be found by setting the partial derivatives of $L_x(P)$ and $L_y(P)$ to zero, i.e.,

$$\frac{\partial L_x(P)}{\partial X} = AX - b_x = 0 \text{ and } \frac{\partial L_y(P)}{\partial Y} = AY - b_y = 0$$

where A is a matrix with $A[i][j] = -c(i,j)$ when $i \neq j$, and $A[i][i] = $ the sum of incident connection weights of cell i. X is a vector of all the x-coordinates of the non-fixed cells, and b_x is a vector with $b_x[i] = $ the sum of x-coordinates of all fixed cells attached to i. Y is a vector of all the y-coordinates of the non-fixed cells, and b_y is a vector with $b_y[i] = $ the sum of y-coordinates of all fixed cells attached to i.

This is a system of linear equations for which iterative numerical methods can be used to find a solution. Known methods include the *conjugate gradient* (*CG*) method [4.27] and the *successive over-relaxation* (*SOR*) method.

During *detailed placement*, cells are spread out to remove all overlaps. Known methods include those described for min-cut placement (Sec. 4.3.1) [4.33] and force-directed placement [4.9].

Example: Quadratic Placement
Given: (1) placement P with two fixed points
p_1 (100,175) and p_2 (200,225), (3) three free
blocks a-c and (4) nets N_1-N_4.

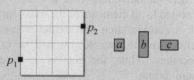

N_1 (P_1,a) N_2 (a,b) N_3 (b,c) N_4 (c,P_2)

Task: find the coordinates of blocks (x_a, y_a), (x_b, y_b) and (x_c, y_c).

Solution:
Solve for x-coordinates.
$$L_x(P) = (100 - x_a)^2 + (x_a - x_b)^2 + (x_b - x_c)^2 + (x_c - 200)^2$$
$$\frac{\partial L_x(P)}{x_a} = -2(100 - x_a) + 2(x_a - x_b) = 4x_a - 2x_b - 200 = 0$$
$$\frac{\partial L_x(P)}{x_b} = -2(x_a - x_b) + 2(x_b - x_c) = -2x_a + 4x_b - 2x_c = 0$$
$$\frac{\partial L_x(P)}{x_c} = -2(x_b - x_c) + 2(x_c - 200) = -2x_b + 4x_c - 400 = 0$$

Put in matrix form $AX = b_x$.
$$\begin{bmatrix} 4 & -2 & 0 \\ -2 & 4 & -2 \\ 0 & -2 & 4 \end{bmatrix} \begin{bmatrix} x_a \\ x_b \\ x_c \end{bmatrix} = \begin{bmatrix} 200 \\ 0 \\ 400 \end{bmatrix} \rightarrow \begin{bmatrix} 2 & -1 & 0 \\ -1 & 2 & -1 \\ 0 & -1 & 2 \end{bmatrix} \begin{bmatrix} x_a \\ x_b \\ x_c \end{bmatrix} = \begin{bmatrix} 100 \\ 0 \\ 200 \end{bmatrix}$$
Solve for X: $x_a = 125$, $x_b = 150$, $x_c = 175$.

Solve for y-coordinates.
$$L_y(P) = (175 - y_a)^2 + (y_a - y_b)^2 + (y_b - y_c)^2 + (y_c - 225)^2$$
$$\frac{\partial L_y(P)}{y_a} = -2(175 - y_a) + 2(y_a - y_b) = 4y_a - 2y_b - 350 = 0$$
$$\frac{\partial L_y(P)}{y_b} = -2(y_a - y_b) + 2(y_b - y_c) = -2y_a + 4y_b - 2y_c = 0$$
$$\frac{\partial L_y(P)}{y_c} = -2(y_b - y_c) + 2(y_c - 225) = -2y_b + 4y_c - 450 = 0$$

Put in matrix form $AY = b_y$.

$$\begin{bmatrix} 4 & -2 & 0 \\ -2 & 4 & -2 \\ 0 & -2 & 4 \end{bmatrix} \begin{bmatrix} y_a \\ y_b \\ y_c \end{bmatrix} = \begin{bmatrix} 350 \\ 0 \\ 450 \end{bmatrix} \rightarrow \begin{bmatrix} 2 & -1 & 0 \\ -1 & 2 & -1 \\ 0 & -1 & 2 \end{bmatrix} \begin{bmatrix} y_a \\ y_b \\ y_c \end{bmatrix} = \begin{bmatrix} 175 \\ 0 \\ 225 \end{bmatrix}$$

Solve for Y: $y_a = 187.5$, $y_b = 200$, $y_c = 212.5$.

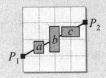

Final solution: a (125,187.5), b (150,200) and c (175,212.5).

Force-directed placement. During force-directed placement, cells and wires are modeled using the mechanical analogy of a mass-spring system, i.e., masses connected to *Hooke's-Law* springs. Each cell exercises attraction toward other cells, where the attraction force is directly proportional to distance. If free movement is granted to all elements of the mass-spring system, then all cells will eventually settle in a configuration that achieves force-equilibrium. Relating this back to circuit placement, the wirelength will be minimized if all cells reach their equilibrium locations. Thus, the goal is to have all cells settle to a placement with force equilibrium.

Force-directed placement, developed in the late 1970s by *N. R. Quinn* [4.24], is a special case of quadratic placement. The potential energy of a stretched Hooke's-Law spring between cells a and b is directly proportional to the squared Euclidean distance between a and b. Furthermore, the spring force exerted on a given cell i is the partial derivative of potential energy with respect to i's position. Thus, determining the energy-minimal positions of cells is equivalent to minimizing the sum of squared Euclidean distances.

Given two connected cells a and b, the attraction force $\overrightarrow{F_{ab}}$ exerted on a by b is

$$\overrightarrow{F_{ab}} = c(a,b) \cdot (\vec{b} - \vec{a})$$

where $c(a,b)$ is the connection weight (priority) between cells a and b, and $(\vec{b} - \vec{a})$ is the vector difference of the positions of a and b in the Euclidean plane. The sum of forces exerted on a cell i connected to other cells j is (Fig. 4.9)

$$\overrightarrow{F_i} = \sum_{c(i,j) \neq 0} \overrightarrow{F_{ij}}$$

The position that minimizes this sum of forces is known as the *zero-force target* (*ZFT*).

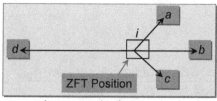

$$\min \overrightarrow{F_i} = c(i,a) \cdot (\overrightarrow{a} - \overrightarrow{i}) + c(i,b) \cdot (\overrightarrow{b} - \overrightarrow{i}) \\ + c(i,c) \cdot (\overrightarrow{c} - \overrightarrow{i}) + c(i,d) \cdot (\overrightarrow{d} - \overrightarrow{i})$$

Fig. 4.9 ZFT position of cell i, which is connected to four other cells a-d.

Using this formulation, two possible extensions are as follows. First, in addition to the attraction force, a repulsion force can be added for unconnected cells to avoid overlap and take into account the overall layout (for a balanced placement). These forces form a system of linear equations that can be solved efficiently using analytic techniques, as with quadratic placement. Second, for each cell, there is an ideal minimum-energy (ZFT) position. By iteratively moving each cell to this position (or nearby, if that position is already occupied) or by cell swapping, gradual improvements can be made. In the end, all cells will be in the position with minimum-force equilibrium configuration.

Basic force-directed placement algorithms. Force-directed placement algorithms iteratively move all cells to their respective ZFT positions. In order to find a cell i's ZFT position (x_i^0, y_i^0), all forces that affect i must be taken into account. Since the ZFT position minimizes the force, both the x- and y- direction forces are set to zero.

$$\sum_{c(i,j)\neq 0} c(i,j) \cdot (x_j^0 - x_i^0) = 0 \quad \text{and} \quad \sum_{c(i,j)\neq 0} c(i,j) \cdot (y_j^0 - y_i^0) = 0$$

Rearranging the variables to solve for x_i^0 and y_i^0 yields

$$x_i^0 = \frac{\sum_{c(i,j)\neq 0} c(i,j) \cdot x_j^0}{\sum_{c(i,j)\neq 0} c(i,j)} \quad \text{and} \quad y_i^0 = \frac{\sum_{c(i,j)\neq 0} c(i,j) \cdot y_j^0}{\sum_{c(i,j)\neq 0} c(i,j)}$$

Using these equations, the ZFT position of a cell i that is connected to other cells j can be computed.

Example: ZFT Position
Given: (1) circuit with NAND gate (cell) a (left), (2) four I/O pads – *In1*-*In3* and *Out* – and their positions, (3) a 3 × 3 layout grid (right), and (4) weighted connections.

In1 (2,2)	*In2* (0,2)	*In3* (0,0) *Out* (2,0)
$c(a,In1) = 8$	$c(a,In2) = 10$	
$c(a,In3) = 2$	$c(a,Out) = 2$	

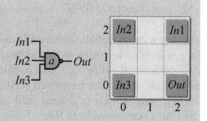

Task: find the ZFT position of cell a.

Solution:

$$x_a^0 = \frac{\sum_{c(i,j)\neq 0} c(a,j)\cdot x_j^0}{\sum_{c(i,j)\neq 0} c(a,j)} = \frac{c(a,In1)\cdot x_{In1} + c(a,In2)\cdot x_{In2} + c(a,In3)\cdot x_{In3} + c(a,Out)\cdot x_{Out}}{c(a,In1) + c(a,In2) + c(a,In3) + c(a,Out)}$$

$$= \frac{8\cdot 2 + 10\cdot 0 + 2\cdot 0 + 2\cdot 2}{8 + 10 + 2 + 2} = \frac{20}{22} \approx 0.9$$

$$y_a^0 = \frac{\sum_{c(i,j)\neq 0} c(a,j)\cdot y_j^0}{\sum_{c(i,j)\neq 0} c(a,j)} = \frac{c(a,In1)\cdot y_{In1} + c(a,In2)\cdot y_{In2} + c(a,In3)\cdot y_{In3} + c(a,Out)\cdot y_{Out}}{c(a,In1) + c(a,In2) + c(a,In3) + c(a,Out)}$$

$$= \frac{8\cdot 2 + 10\cdot 2 + 2\cdot 0 + 2\cdot 0}{8 + 10 + 2 + 2} = \frac{36}{22} \approx 1.6$$

ZFT position of cell a is at (1,2).

The following gives a high-level sketch of the force-directed placement approach.

Force-Directed Placement Algorithm
Input: set of all cells V
Output: placement P

```
1.  P = PLACE(V)                        // arbitrary initial placement
2.  loc = LOCATIONS(P)                  // set coordinates for each cell in P
3.  foreach (cell c ∈ V)
4.      status[c] = UNMOVED
5.  while (ALL_MOVED(V) || !STOP())      // continue until all cells have been
                                         //    moved or some stopping
                                         //    criterion is reached
6.      c = MAX_DEGREE(V,status)         // unmoved cell that has largest
                                         //    number of connections
7.      ZFT_pos = ZFT_POSITION(c)        // ZFT position of c
8.      if (loc[ZFT_pos] == Ø)           // if position is unoccupied,
9.          loc[ZFT_pos] = c             //    move c to its ZFT position
10.     else
11.         RELOCATE(c,loc)              // use methods discussed below
12.     status[c] = MOVED                // mark c as moved
```

Starting with an initial placement (line 1), the algorithm selects a cell c that has the largest number of connections and has not been moved (line 6). It then calculates the ZFT position of c (line 7). If that position is unoccupied, then c is moved there (lines 8-9). Otherwise, c can be moved according to several methods discussed next (lines 10-11). This process continues until some stopping criterion is reached or until all cells have been considered (lines 5-12).

Finding a valid location for a cell with an occupied ZFT position. Let p be an incoming cell and let q be the cell that is currently in p's ZFT position. Following are four options for how to rearrange p and q.

− If possible, move p to a cell position close to q.
− Compute the cost difference if p and q were to be swapped. If the total cost reduces, i.e., the weighted connection length $L(P)$ is smaller, then swap p and q.
− *Chain move*: cell p is moved to cells q's location. Cell q, in turn, is shifted to the next position. If a cell r is occupying this space, cell r is shifted to the next position. This continues until all affected cells are placed.
− *Ripple move*: cell p is moved to cell q's location. A new ZFT position for cell q is computed (discussed below). This ripple effect continues until all cells are placed.

Example: Force-Directed Placement
Given: (1) placement of blocks b_1-b_3 (right) and (2) weighted nets N_1 and N_2.

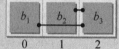

$N_1 = (b_1,b_3)$ $N_2 = (b_2,b_3)$ $c(N_1) = 2$ $c(N_2) = 1$

Task: use force-directed placement to find a heuristic minimum-wirelength placement.

Solution:
Consider cell b_3.

ZFT position for $b_3 = x^0_{b_3} = \dfrac{\sum\limits_{c(b_3,j)\neq 0} c(b_3,j) \cdot x^0_j}{\sum\limits_{c(b_3,j)\neq 0} c(b_3,j)} = \dfrac{2 \cdot 0 + 1 \cdot 1}{2 + 1} \approx 0$, which is occupied by cell b_1.

$L(P)$ before move = 3, $L(P)$ after move = 3. Cells b_3 and b_1 should not swap.

Consider cell b_2.

ZFT position for $b_2 = x^0_{b_2} = \dfrac{\sum\limits_{c(b_2,j)\neq 0} c(b_2,j) \cdot x^0_j}{\sum\limits_{c(b_2,j)\neq 0} c(b_2,j)} = \dfrac{1 \cdot 2}{1} = 2$, which is occupied by cell b_3.

$L(P)$ before move = 3, $L(P)$ after move = 2. Cells b_2 and b_3 should swap.

Force-directed placement with ripple moves. In force-directed placement with ripple moves [4.30], the cells are sorted in descending order according to their connection degrees (lines 1-3). During each iteration (line 4), the ZFT position for the next cell in the list (*seed*) is computed (line 9). If the position is free, *seed* is moved there (lines 12-17). If the position is occupied, then its current inhabitant is moved next, assuming it has not already been moved. In order to avoid infinite loops, once a cell has been moved, it is marked as *LOCKED* and can no longer be moved until the next iteration (lines 18-40).

Force-Directed Placement with Ripple Moves Algorithm
Input: set of all cells *V*
Output: positions of each cell *pos*
1.   **foreach** (cell $c \in V$)
2.   *degree[c]* = CONNECTION_DEGREE(*c*)   // compute connection degree
3.   *L* = SORT(*V,degree*)   // descending order by degree
4.   **while** (*iteration_count* < *iteration_limit*)
5.   *end_ripple_move* = **false**
6.   *seed* = NEXT_ELEMENT(*L*)   // next cell in *L*
7.   *pos_type[seed]* = *VACANT*   // position type = *VACANT*
8.   **while** (*end_ripple_move* == **false**)
9.   *curr_pos* = ZFT_POSITION(*seed*)
10.   *curr_pos_type* = ZFT_POSITION_TYPE(*seed*)
11.   **switch** (*curr_pos_type*)
12.   **case** *VACANT*:
13.   *pos[seed]* = MOVE(*seed,curr_pos*)   // move *seed* to *curr_pos*
14.   *pos_type[seed]* = *LOCKED*
15.   *end_ripple_move* = **true**
16.   *abort_count* = 0
17.   **break**
18.   **case** *LOCKED*:
19.   *pos[seed]* = MOVE(*seed,*NEXT_FREE_POS())
20.   *pos_type[seed]* = *LOCKED*
21.   *end_ripple_move* = **true**
22.   *abort_count* = *abort_count* + 1
23.   **if** (*abort_count* > *abort_limit*)
24.   **foreach** (*pos_type[c]* == *LOCKED*)
25.   RESET(*pos_type[c]*)
26.   *iteration_count* = *iteration_count* + 1
27.   **break**
28.   **case** *SAME_AS_PRESENT_LOCATION*:
29.   *pos[seed]* = MOVE(*seed,curr_pos*)
30.   *pos_type[seed]* = *LOCKED*
31.   *end_ripple_move* = **true**
32.   *abort_count* = 0
33.   **break**
34.   **case** *OCCUPIED*:   // occupied but not locked
35.   *prev_cell* = CELL(*curr_pos*)   // cell in ZFT position
36.   *pos[seed]* = MOVE(*seed,curr_pos*)
37.   *pos_type[seed]* = *LOCKED*
38.   *seed* = *prev_cell*
39.   *end_ripple_move* = **false**
40.   *abort_count* = 0
41.   **break**

In each iteration, an upper bound of fixed cells (*abort_limit*) is imposed. If the number of fixed cells exceeds this limit, then all fixed cells are released and the cell with the next highest connectivity is considered in the new iteration (lines 23-26). Each cell has one of four options (lines 11-41).

− *VACANT*: ZFT position is free. Place the cell there and proceed with the next iteration (lines 12-17).
− *LOCKED*: ZFT position is occupied and fixed. Place the cell at the next free position and increment *abort_count*. If *abort_count* > *abort_limit*, stop the ripple moves and go on with the next cell in list L (lines 18-27).
− *SAME_AS_PRESENT_LOCATION*: ZFT position is the same as the current position of the cell. Place cell here and consider the next cell in L (lines 28-33).
− *OCCUPIED*: ZFT position is occupied but not by a locked cell. Place the cell here and move the cell that had occupied the ZFT position (lines 34-41).

The inner while loop (cell movement) is called only when *end_ripple_move* is **false**. Note that this flag is set to **true** when the seed cell's ZFT position is *VACANT*, *LOCKED* or *SAME_AS_PRESENT_LOCATION*. In these cases, the algorithm can move on with the next cell in the list. The outer while loop goes through list L and finds the best ZFT position of each cell until *iteration_limit* is reached.

▶ 4.3.3 Simulated Annealing

Simulated annealing (SA) (Sec. 3.5.3) is the basis of some of the most well-known placement algorithms.

```
Simulated Annealing Algorithm for Placement
Input: set of all cells V
Output: placement P
1.   T = T₀                              // set initial temperature
2.   P = PLACE(V)                        // arbitrary initial placement
3.   while (T > Tmin)
4.     while (!STOP())                   // not yet in equilibrium at T
5.       new_P = PERTURB(P)
6.       Δcost = COST(new_P) – COST(P)
7.       if (Δcost < 0)                  // cost improvement
8.         P = new_P                     // accept new placement
9.       else                           // no cost improvement
10.        r = RANDOM(0,1)               // random number [0,1)
11.        if (r < e^(-Δcost/T))         // probabilistically accept
12.          P = new_P
13.    T = α · T                         // reduce T, 0 < α < 1
```

From an initial placement (line 2), the PERTURB function generates a new placement *new_P* by perturbing the current placement (line 5). $\Delta cost$ records the cost difference between the previous placement P and *new_P* (line 6). If the cost improves, i.e., $\Delta cost < 0$, then *new_P* is accepted (lines 7-8). Otherwise, *new_P* is probabilistically accepted (lines 9-12). Once the annealing process has "reached equilibrium" (e.g., after a prescribed number of move attempts) at the current temperature, the temperature is decreased. This process continues until $T < T_{min}$.

TimberWolf. One of the earliest academic packages, *TimberWolf*, was developed at the *University of California, Berkeley* by *C. Sechen* and later commercialized [4.28] [4.29]. From the widths of all cells in the netlist, the algorithm produces an initial placement with all cells in rows, along with a target row length. TimberWolf allots additional area around any macro cells to allow sufficient interconnect area, and uses a specialized pin assignment algorithm to optimize any macro-cell pins that are not fixed to specific locations. The macro cells are then placed simultaneously along with the standard cells. The original TimberWolf (v.3.2) only optimizes the placement of standard cells while I/Os and macros stay in their original locations.

The placement algorithm consists of the following three stages.

1. Place standard cells with minimum total wirelength using simulated annealing.
2. Globally route (Chap. 5) the placement by introducing routing channels, when necessary. Recompute and minimize the total wirelength.
3. Locally optimize the placement with the goal of minimizing channel height.

The remainder of this section discusses the first (placement) stage. For further reading on TimberWolf, see [4.29].

Perturb. The PERTURB function generates a new placement from an existing placement using one of the following actions.

—   MOVE: Shift a cell to a new position (another row)
—   SWAP: Exchange two cells
—   MIRROR: Reflect the cell orientation around the *y*-axis, used only if MOVE
                  and SWAP are infeasible

The scope of PERTURB is limited to a small window of size $w_T \times h_T$ (Fig. 4.10). For MOVE, a cell can only be moved within this window.

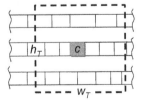

Fig. 4.10 Window with the dimensions $w_T \times h_T$ around a standard cell c.

For SWAP, two cells a (x_a, y_a) and b (x_b, y_b) can be exchanged only if

$$|x_a - x_b| \leq w_T \quad \text{and} \quad |y_a - y_b| \leq h_T$$

The window size (w_T, h_T) depends on the current temperature T, and decreases as temperature reduces. The window size for the next iteration is based on the current temperature T_{curr} and the next iteration's temperature T_{next} using

$$w_{T_{next}} = w_{T_{curr}} \frac{\log(T_{next})}{\log(T_{curr})} \quad \text{and} \quad h_{T_{next}} = h_{T_{curr}} \frac{\log(T_{next})}{\log(T_{curr})}$$

Cost. The COST function in TimberWolf (v3.2) is defined as $\Gamma = \Gamma_1 + \Gamma_2 + \Gamma_3$, the sum of three parameters – (1) total estimated wirelength Γ_1, (2) amount of overlap Γ_2, and (3) row inequality length Γ_3.

Γ_1 is computed as the summation of each net's *half-perimeter wirelength (HPWL)*, which is defined as its horizontal span plus its vertical span. Weights for each direction, horizontal weight w_H and vertical weight w_V, can also be applied. Given a priority weight γ_1, Γ_1 is defined as the sum of the total wirelength over all nets *net* $\in$ *Netlist*, where *Netlist* is the set of all nets.

$$\Gamma_1 = \gamma_1 \cdot \sum_{net \in Netlist} w_H(net) \cdot x_{net} + w_V(net) \cdot y_{net}$$

A higher weight value for *net* gives higher emphasis on reducing *net*'s wirelength. Weights can also be used for direction control – giving preference to a certain wiring direction. During standard-cell placement where feedthrough cells are limited, low horizontal weights $w_H(net)$ encourage the usage of horizontal channels rather than the vertical connections.

Γ_2 represents the total cell overlap of the placement. Let $o(i,j)$ represent the area of overlap between cells i and j. Given a priority weight γ_2, Γ_2 is defined as the sum of the square of all cell overlaps between cells i and j, where $i \in V, j \in V, i \neq j$, with V being the set of all cells.

$$\Gamma_2 = \gamma_2 \cdot \sum_{i \in V, j \in V, i \neq j} o(i, j)^2$$

Larger overlaps, which require more effort to correct, are penalized more heavily due to the quadratic form.

Γ_3 represents the cost of all row lengths $L(row)$ that deviate from the goal length $L_{opt}(row)$ during placement. Cell movement can often lead to row length variation, where the resulting rows lengths deviate from the goal length. In practice, uneven rows can waste area and induce uneven wire distributions. Both phenomena can lead to increased total wirelength and total congestion. Given a priority factor γ_3, Γ_3 is defined as the sum of row length deviation for all rows *row* $\in$ *Rows*, where *Rows* is the set of all rows.

$$\Gamma_3 = \gamma_3 \cdot \sum_{row \in Rows} |L(row) - L_{opt}(row)|$$

Temperature Reduction. The temperature T is reduced by a cooling factor α. This value is empirically chosen and often depends on the temperature range. The annealing process starts at a high temperature, such as $4 \cdot 10^6$ (units do not play a role). Initially, the temperature is reduced quickly ($\alpha \approx 0.8$). After a certain number of iterations, the temperature reduces at a slower rate ($\alpha \approx 0.95$), when the placement is being fine-tuned. Toward the end, the temperature is again reduced at a fast pace ($\alpha \approx 0.8$), corresponding to a "quenching" step. TimberWolf finishes when $T < T_{min}$ where $T_{min} = 1$.

Number of Times Through the Inner Loop. At each temperature, a number of calls are made to PERTURB to generate new placements. This number is intended to achieve *equilibrium* at the given temperature, and depends on the size of the design. The authors of [4.29] experimentally determined that designs with ~200 cells require 100 iterations per cell, or roughly $2 \cdot 10^4$ runs per temperature step. Other simulated annealing approaches use acceptance ratio as an equilibrium criterion, e.g., *Lam* [4.20] shows that a target acceptance ratio of 44% produces competitive results.

▶ **4.3.4 Modern Placement Algorithms**

Algorithms for global placement have been studied by many researchers since the late 1980s, and the prevailing paradigm has changed several times to address new challenges arising in commercial chip designs [4.6][4.22]. This section reviews modern algorithms for global placement, while the next section covers legalization and detailed placement, as well as the need for such a separation of concerns. Timing-driven placement is discussed in Sec. 8.3.

The global placement algorithms in use today can handle extremely large netlists using analytic techniques, i.e., by modeling interconnect length with mathematical functions and optimizing these functions with numerical methods. Dimensions and sizes of standard cells are initially ignored to quickly find a seed placement, but are then gradually factored into the placement optimization so as to avoid uneven densities or routing congestion. Two common paradigms are based on *quadratic* and *force-directed placement*, and on *nonlinear optimization*. The former was introduced earlier and seeks to approximate wirelength by quadratic functions, which can be minimized by solving linear systems of equations (Sec. 4.3.2). The latter relies on more sophisticated functions to approximate interconnect length, and requires more sophisticated numerical optimization algorithms [4.4][4.16][4.17].

Of the two types, quadratic methods are easier to implement and appear to be more scalable in terms of runtime. Nonlinear methods require careful tuning to achieve numerical stability and often run much slower than quadratic methods. However, nonlinear methods can better account for the shapes and sizes of standard cells and especially macro blocks, whereas quadratic placement requires dedicated *spreading* techniques. Both placement techniques are often combined with netlist clustering to reduce runtime [4.4][4.16][4.17][4.34], in a manner that is conceptually similar to multilevel partitioning (Chap. 2). However, the use of clustering in placement often leads to losses in solution quality. Thus, it is an open question whether the multilevel approach can outperform the flat approach in terms of runtime and solution quality for placement, as is the case for partitioning [4.5].

The aspects of quadratic placement that appear most impactful in practice are (1) the representation of multi-pin nets by sets of graph edges (net modeling), (2) the choice of algorithms for spreading, and (3) the strategy for interleaving spreading with quadratic optimization. Two common net models include *cliques*, where every pair of pins is connected by an edge with a small weight, and *stars*, where every pin is connected to a "star-point" that represents the net (or hyperedge) itself [4.12]. Edges representing a net are given fractional weights that add up to the net's weight (or to unity). For nets with fewer pins, cliques are preferred because they do not introduce new variables. For larger nets, stars are useful because they entail only a linear number of graph edges [4.34]. The star-point can be movable or placed in the centroid (average location or *barycenter*) of its neighbors. The latter option is preferred in practice because (1) it corresponds to the optimal location of the star-point in quadratic placement, and (2) it saves two (x,y) variables. Some placers additionally use a *linearization* technique that assigns a constant weight

$$w(i, j) = \frac{1}{x_i - x_j}$$

to each quadratic term $(x_i - x_j)^2$ within the objective function. The weight $w(i,j)$ has the effect of turning each such *squared wirelength term* into a *linear wirelength term*, and can therefore be truer to an underlying linear wirelength objective. These weights are treated as constants, and then updated between rounds of quadratic optimization. A more accurate placement-dependent net model is proposed in [4.32].

Spreading is based on estimates of cell density in different regions of the chip. These estimates are computed by allocating movable objects into bins of a regular grid, and comparing their total area to available capacity per bin. Spreading can be performed after quadratic optimization using a combination of sorting by location and geometric scaling [4.32]. For example, cells in a dense region may be sorted by their x-coordinates and then re-placed in this order, so as to avoid overlaps. An implicit spreading method to reduce overlap is to enclose a set of cells in a rectangular region and then perform linear scaling [4.18].

Spreading can also be integrated directly into quadratic optimization by adding *spreading forces* that push movable objects away from dense regions. These additional forces are modeled by imaginary fixed pins (anchor points) and imaginary wires pulling individual standard cells toward fixed pins [4.11]. This integration allows conventional quadratic placement to trade interconnect minimization for smaller overlaps between modules. *FastPlace* [4.34] first performs simple geometric scaling and then uses the resulting locations as anchor points during quadratic optimization. These steps of spreading and quadratic optimization are interleaved in FastPlace to encourage spreading that does not conflict with interconnect optimization. Researchers have also sought to develop spreading algorithms that are sufficiently accurate to be invoked only once after quadratic optimization [4.35].

Analytic placement can be extended to optimize not only interconnect length, but also routing congestion [4.31]. This requires wiring congestion estimation, which is similar to density estimation and is also maintained on a regular grid. Congestion information can be used in the same ways as density estimation to perform spreading. Some researchers have developed post-processors to improve congestion properties of a given placement [4.21].

Several modern placers are available free of charge for research purposes. As of 2010, the most accessible placers are *APlace* [4.16][4.17], *Capo* [4.2][4.26], *FastPlace 3.0* [4.34], *mPL6* [4.4], and *simPL* [4.18]. All except simPL[3] are equipped with legalizers and detailed placers so as to produce legal and highly optimized solutions. mPL6 is significantly slower than FastPlace, but finds solutions with smaller total interconnect length. Capo, a min-cut placer, is available in C++ source code. Its runtime is between that of mPL6 and FastPlace, but in many cases it produces solutions that are inferior to FastPlace solutions in terms of total interconnect length. However, for designs where achieving routability is difficult, Capo offers a better chance to produce a routable placement. It is also competitive on smaller designs (below 50,000 movable objects), especially those with high density and many fixed obstacles.

4.4   4.4 Legalization and Detailed Placement

Global placement assigns locations to standard cells and larger circuit modules, e.g., macro blocks. However, these locations typically do not align with power rails, and may have continuous (real) coordinates rather than discrete coordinates. Therefore, the global placement must be *legalized*. The allowed legal locations are equally spaced within pre-defined rows, and the point-locations from global placement should be snapped to the closest possible legal locations (Fig. 4.11).

Legalization is necessary not only *after global placement*, but also *after incremental changes* such as cell resizing and buffer insertion during physical synthesis (Sec. 8.5). Legalization seeks to find legal, non-overlapping placements for all placeable modules so as to minimize any adverse impact on wirelength, timing and other design objectives. Unlike algorithms for "cell spreading" during global placement (Sec. 4.3), legalization typically assumes that the cells are distributed fairly well

---

[3] simPL uses FastPlace-DP [4.23] for both legalization and detailed placement.

throughout the layout region and have relatively small mutual overlap. Once a legal placement is available, it can be improved with respect to a given objective by means of *detailed placement* techniques, such as swapping neighboring cells to reduce total wirelength, or sliding cells to one side of the row when unused space is available. Some detailed placers target routability, given that route topologies can be determined once a legal placement is available.

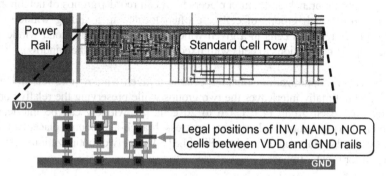

Fig. 4.11 An instance of detailed placement. Between each *VDD* and *GND* rail, cells must be placed in non-overlapping site-aligned locations.

One of the simplest and fastest techniques for legalization is *Tetris* by *D. Hill* [4.10]. It sorts cells by x-coordinate and processes them greedily. Each cell is placed at the closest available legal location that does not exceed the row capacity. In its purest form, the Tetris algorithm has several known drawbacks, one being its ignorance of the netlist. Another drawback is that in the presence of large amounts of whitespace, the Tetris algorithm can place cells onto one side of the layout, and thus considerably increase wirelength or circuit delay if some I/O pads are fixed on the opposite side of the layout.

Several variations of the Tetris algorithm exist. One variant subdivides the layout into regions and performs Tetris legalization in each region: this prevents the large displacements observed with the original version. Other variants try to account for wirelength while finding the best legal location for a given cell [4.10]. There are also versions that inject unused space between legalized cells, so as to avoid routing congestion [4.21].

Some algorithms for legalization and placement are co-developed with global placement algorithms. For instance, in the context of min-cut placement, detailed placement can be performed by optimal partitioners and end-case placers [4.1][4.3] invoked in very small (end-case) bins that are produced after the netlist is repeatedly partitioned. Given that these end-case bins contain a very small number of cells (four to six), optimal locations can be found by exhaustive enumeration or *branch-and-bound*. For larger bins, partitioning can be performed optimally (with up to ~35 cells in a bin). Some analytic algorithms perform legalization in iterations [4.32]. At each iteration, cells closest to legal sites are identified and snapped to legal sites, then considered fixed thereafter. After a round of analytic placement, another group of cells is snapped to legal sites, and the process continues until all cells have been given legal locations.

A common problem with simple and fast legalization algorithms is that some cells may travel a long distance, thus significantly increasing the length and, hence, delay of incident nets. This phenomenon can be mitigated by detailed placement. For example, optimal branch-and-bound placers [4.3] can reorder groups of neighboring cells in a row. Such groups of cells are often located in a sliding window; the optimal placer reorders cells in a given window so as to improve total wirelength (accounting for connections to cells with fixed locations outside the window).

A more scalable optimization splits the cells in a given window into left and right halves, and optimally interleaves the two groups while preserving the relative order of cells from each group [4.12]. Up to 20 cells per window can be interleaved efficiently during detailed placement, whereas branch-and-bound placement can typically handle only up to eight cells [4.3]. These two optimizations can be combined for greater impact.

Sometimes, wirelength can be improved by reordering cells that are not adjacent. For example, pairs of non-adjacent cells connected by a net can be swapped [4.23], and sets of three such cells can be cycled. Yet another detailed placement optimization is possible when unused space is available between cells placed in a row. These cells can be shifted to either side, or to intermediate locations. Optimal locations to minimize wirelength can be found by a polynomial-time algorithm [4.15], which is practical in many applications.

Software implementations of legalization and detailed placement are often bundled, but are sometimes independent of global placement. One example is *FastPlace-DP* [4.23] (binary available from the authors). FastPlace-DP works best when the input placement is almost legal or requires only a small number of local changes. FastPlace-DP performs a series of simple but efficient incremental optimizations which typically decrease interconnect length by several percent. On the other end of the spectrum is *ECO-System* [4.25]. It is integrated with the *Capo* placer [4.2][4.26] and uses more sophisticated yet slower optimizations. ECO-System first analyzes a given placement and identifies regions where cells overlap so much that they need to be re-placed. The Capo algorithm is then applied simultaneously to each region so as to ensure consistency. Capo integrates legalization and detailed placement into global min-cut placement. Therefore, ECO-System will produce a legal placement even if the initial placement requires significant changes.

Other strategies, such as the use of *linear programming* [4.7] and *dynamic programming* [4.14], have been integrated into legalization and detailed placement with promising results. The legalization of mixed-size netlists that contain large movable blocks is particularly challenging [4.14].

Chapter 4 Exercises

Exercise 1: Estimating Total Wirelength

Consider the five-pin net with pins *a-e* (right). Each grid edge has unit length.

(a) Draw a rectilinear minimum-length chain, a rectilinear minimum spanning tree (RMST), and a rectilinear Steiner minimum tree (RSMT) to connect all pins.

(b) Find the weighted total wirelength using each estimation technique from (a) if each grid edge has weight = 2.

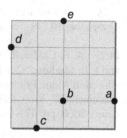

Exercise 2: Min-Cut Placement

Perform min-cut placement to place gates *a-g* on a 2 × 4 grid. Use the Kernighan-Lin algorithm for partitioning. Use alternating (horizontal and vertical) cutlines. The cutline cut_1 represents the initial vertical cut. Each edge on the grid has capacity $\sigma_P(e) = 2$. Estimate whether the placement is routable.

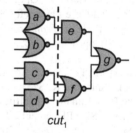

Exercise 3: Force-Directed Placement

A circuit with two gates *a* and *b* and three I/O pads *In*1 (0,2), *In*2 (0,0) and *Out* (2,1) is given (left). The weights of the connections are shown below. Calculate the ZFT positions of the two gates. Place the circuit on a 3 × 3 grid (right).

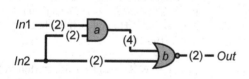

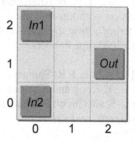

Exercise 4: Global and Detailed Placement

What are the main differences between global and detailed placement? Explain why the global and detailed placement steps are performed separately. Explain why detailed placement follows global placement.

Chapter 4 References

[4.1]   M. Breuer, "Min-Cut Placement", *J. Design Autom. and Fault-Tolerant Computing* 10 (1977), pp. 343-382.

[4.2]   A. E. Caldwell, A. B. Kahng and I. L. Markov, "Can Recursive Bisection Alone Produce Routable Placements?", *Proc. Design Autom. Conf.*, 2000, pp. 477-482.

[4.3]   A. E. Caldwell, A. B. Kahng and I. L. Markov, "Optimal Partitioners and End-Case Placers for Standard-Cell Layout", *IEEE Trans. on CAD* 19(11) (2000), pp. 1304-1313.

[4.4]   T. F. Chan, J. Cong, J. R. Shinnerl, K. Sze and M. Xie, "mPL6: Enhanced Multilevel Mixed-Size Placement", *Proc. Intl. Symp. on Phys. Design*, 2006, pp. 212-221.

[4.5]   H. Chen, C.-K. Cheng, N.-C. Chou, A. B. Kahng, J. F. MacDonald, P. Suaris, B. Yao and Z. Zhu, "An Algebraic Multigrid Solver for Analytical Placement with Layout Based Clustering", *Proc. Design Autom. Conf.*, 2003, pp. 794-799.

[4.6]   J. Cong, J. R. Shinnerl, M. Xie, T. Kong and X. Yuan, "Large-Scale Circuit Placement", *ACM Trans. on Design Autom. of Electronic Sys.* 10(2) (2005), pp. 389-430.

[4.7]   J. Cong and M. Xie, "A Robust Detailed Placement for Mixed-Size IC Designs", *Proc. Asia and South Pacific Design Autom. Conf.*, 2006, pp. 188-194.

[4.8]   A. E. Dunlop and B. W. Kernighan, "A Procedure for Placement of Standard-Cell VLSI Circuits", *IEEE Trans. on CAD* 4(1) (1985), pp. 92-98.

[4.9]   H. Eisenmann and F. Johannes, "Generic Global Placement and Floorplanning", *Proc. Design Autom. Conf.*, 1998, pp. 269-274.

[4.10] D. Hill, *Method and System for High Speed Detailed Placement of Cells Within an Integrated Circuit Design*, U.S. Patent 6370673, 2001.

[4.11] B. Hu, Y. Zeng and M. Marek-Sadowska, "mFAR: Fixed-Points-Addition-Based VLSI Placement Algorithm", *Proc. Intl. Symp. on Phys. Design*, 2005, pp. 239-241.

[4.12] T. C. Hu and K. Moerder, "Multiterminal Flows in a Hypergraph", in *VLSI Circuit Layout: Theory and Design* (T. C. Hu and E. S. Kuh, eds.), IEEE, 1985.

[4.13] S. W. Hur and J. Lillis, "Mongrel: Hybrid Techniques for Standard Cell Placement", *Proc. Intl. Conf. on CAD*, 2000, pp. 165-170.

[4.14] A. B. Kahng, I. L. Markov and S. Reda, "On Legalization of Row-Based Placements", *Proc. Great Lakes Symp. on VLSI*, 2004, pp. 214-219.

[4.15] A. B. Kahng, P. Tucker and A. Zelikovsky, "Optimization of Linear Placements for Wirelength Minimization with Free Sites", *Proc. Asia and South Pacific Design Autom. Conf.*, 1999, pp. 241-244.

[4.16] A. B. Kahng and Q. Wang, "Implementation and Extensibility of an Analytic Placer", *IEEE Trans. on CAD* 24(5) (2005), pp. 734-747.

[4.17] A. B. Kahng and Q. Wang, "A Faster Implementation of APlace", *Proc. Intl. Symp. on Phys. Design*, 2006, pp. 218-220.

[4.18] M.-C. Kim, D.-J. Lee and I. L. Markov, "simPL: An Effective Placement Algorithm", *Proc. Intl. Conf. on CAD*, 2010.

[4.19] J. B. Kruskal, "On the Shortest Spanning Subtree of a Graph and the Traveling Salesman Problem", *Proc. Amer. Math. Soc.* 7(1) (1956), pp. 8-50.

[4.20] J. K. Lam, *An Efficient Simulated Annealing Schedule* (*Doctoral Dissertation*), Yale University, 1988.

[4.21] C. Li, M. Xie, C.-K. Koh, J. Cong and P. H. Madden, "Routability-Driven Placement and White Space Allocation", *IEEE Trans. on CAD* 26(5) (2007), pp. 858-871.

[4.22] G.-J. Nam and J. Cong, eds., *Modern Circuit Placement: Best Practices and Results*, Springer, 2007.

[4.23] M. Pan, N. Viswanathan and C. Chu, "An Efficient and Effective Detailed Placement Algorithm", *Proc. Intl. Conf. on CAD*, 2005, pp. 48-55.

[4.24] N. R. Quinn, "The Placement Problem as Viewed from the Physics of Classical Mechanics", *Proc. Design Autom. Conf.*, 1975, pp. 173-178.

[4.25] J. A. Roy and I. L. Markov, "ECO-System: Embracing the Change in Placement," *IEEE Trans. on CAD* 26(12) (2007), pp. 2173-2185.

[4.26] J. A. Roy, D. A. Papa, S. N. Adya, H. H. Chan, A. N. Ng, J. F. Lu and I. L. Markov, "Capo: Robust and Scalable Open-Source Min-Cut Floorplacer", *Proc. Intl. Symp. on Phys. Design*, 2005, pp. 224-226, vlsicad.eecs.umich.edu/BK/PDtools/.

[4.27] Y. Saad, *Iterative Methods for Sparse Linear Systems*, Soc. of Industrial and App. Math., 2003.

[4.28] C. Sechen, "Chip-Planning, Placement and Global Routing of Macro/Custom Cell Integrated Circuits Using Simulated Annealing", *Proc. Design Autom. Conf.*, 1988, pp. 73-80.

[4.29] C. Sechen and A. Sangiovanni-Vincentelli, "TimberWolf 3.2: A New Standard Cell Placement and Global Routing Package", *Proc. Design Autom. Conf.*, 1986, pp. 432-439.

[4.30] K. Shahookar and P. Mazumder, "VLSI Cell Placement Techniques", *ACM Computing Surveys* 23(2) (1991), pp. 143-220.

[4.31] P. Spindler and F. M. Johannes, "Fast and Accurate Routing Demand Estimation for Efficient Routability-Driven Placement", *Proc. Design, Autom. and Test in Europe*, 2007, pp. 1226-1231.

[4.32] P. Spindler, U. Schlichtmann and F. M. Johannes, "Kraftwerk2 – A Fast Force-Directed Quadratic Placement Approach Using an Accurate Net Model", *IEEE Trans. on CAD* 27(8) (2008), pp. 1398-1411.

[4.33] R.-S. Tsay, E. S. Kuh and C.-P. Hsu, "PROUD: A Sea-of-Gates Placement Algorithm", *IEEE Design and Test* 5(6) (1988), pp. 44-56.

[4.34] N. Viswanathan, M. Pan and C. Chu, "FastPlace 3.0: A Fast Multi-level Quadratic Placement Algorithm with Placement Congestion Control", *Proc. Asia and South Pacific Design Autom. Conf.*, 2007, pp. 135-140.

[4.35] Z. Xiu and R. A. Rutenbar, "Mixed-Size Placement With Fixed Macrocells Using Grid-Warping", *Proc. Intl. Symp. on Phys. Design*, 2007, pp. 103-110.

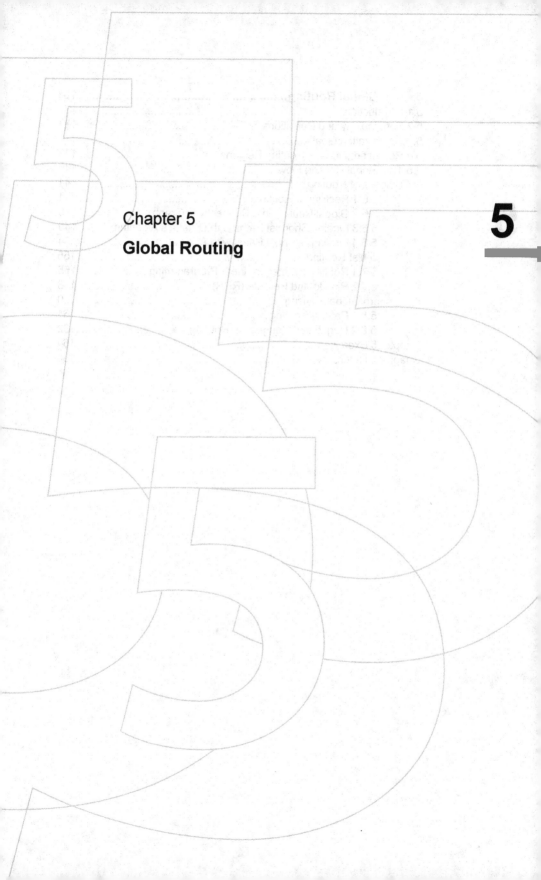

Chapter 5

Global Routing

5

5　　Global Routing..**131**

5

5   Global Routing

During *global routing*, pins with the same electric potential are connected using wire segments. Specifically, after *placement* (Chap. 4), the layout area is represented as *routing regions* (Sec. 5.4) and all nets in the *netlist* are routed in a systematic manner (Sec. 5.5). To minimize total routed length, or optimize other objectives (Sec. 5.3), the route of each net should be short (Sec. 5.6). However, these routes often compete for the same set of limited resources. Such conflicts can be resolved by concurrent routing of all nets (Sec. 5.7), e.g., *integer linear programming* (*ILP*), or by sequential routing techniques, e.g., *rip-up and reroute*. Several algorithmic techniques enable scalability of modern global routers (Sec. 5.8).

5.1 Introduction

A *net* is a set of two or more *pins* that have the same electric potential. In the final chip design, they must be connected. A typical p-pin net connects one output pin of a gate and $p-1$ input pins of other gates; its *fanout* is equal to $p-1$. The term *netlist* refers collectively to all nets.

Given a placement and a netlist, determine the necessary wiring, e.g., net topologies and specific routing segments, to connect these cells while respecting constraints, e.g., design rules and routing resource capacities, and optimizing routing objectives, e.g., minimizing total wirelength and maximizing timing slack.

In area-limited designs, standard cells may be packed densely without unused space. This often leads to routing *congestion*, where the shortest routes of several nets are incompatible because they traverse the same tracks. Congestion forces some routes to detour; thus, in congested regions, it can be difficult to predict the eventual length of wire segments. However, the total wirelength cannot exceed the available routing resources, and in some cases the chip area must be increased to ensure successful routing. *Fixed-die routing*, where the chip outline and all routing resources are fixed, is distinguished from *variable-die routing*, where new routing tracks can be added as needed. For the fixed-die routing problem,[1] 100% routing completion is not always possible *a priori*, but may be possible after changes to placement. On the other hand, in older standard-cell circuits with two or three metal layers, new tracks can be inserted as needed, resulting in the classical variable-die channel routing problem for which 100% routing completion is always possible. Fig. 5.1 outlines the major categories of routing algorithms discussed in this book.

---

[1] The *fixed-die* routing problem is so named because the layout bounding box and the number of routing tracks are predetermined due to the fixed floorplan and power-ground distribution.

MULTI-STAGE ROUTING OF SIGNAL NETS				
GLOBAL ROUTING	**DETAILED ROUTING**	**TIMING-DRIVEN ROUTING**	**LARGE SINGLE-NET ROUTING**	**GEOMETRIC TECHNIQUES**
Coarse-grain assignment of routes to routing regions	Fine-grain assignment of routes to routing tracks	Net topology optimization and resource allocation to critical nets	Power (*VDD*) and Ground (*GND*) routing	Non-Manhattan and clock routing
(Chap. 5)	(Chap. 6)	(Chap. 8)	(Chap. 3)	(Chap. 7)

Fig. 5.1 Routing problem types and the chapters in which they are discussed.

With the scale of modern designs at millions of nets, *global routing* has become a major computational challenge. Full-chip routing is usually performed in three steps: (high-level) *global routing*, (low-level) *detailed routing*, and *timing-driven routing*. The first two steps are illustrated in Fig. 5.2; the last step is discussed in Sec. 8.4.

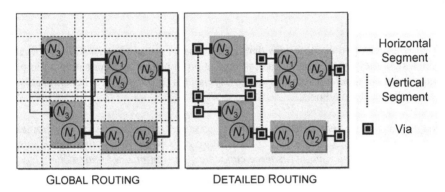

GLOBAL ROUTING   DETAILED ROUTING

Fig. 5.2 Graphical representation of nets N_1-N_3 that are (coarsely) globally routed using *routing regions* (left), and then (finely) detailed routed using rectilinear paths (right). This example assumes two-layer routing, with horizontal and vertical segments routed on separate layers.

During global routing, the wire segments used by net topologies are tentatively assigned (embedded) within the chip layout. The chip area is represented by a coarse routing grid, and available routing resources are represented by edges with capacities in a grid graph. Nets are then assigned to these routing resources.

During detailed routing, the wire segments are assigned to specific routing tracks. This process involves a number of intermediate tasks and decisions such as *net ordering*, i.e., which nets should be routed first, and *pin ordering*, i.e., within a net, in what order should the pins be connected. These two issues are the major challenges in *sequential* routing, where nets are routed one at a time. The net and pin orderings can have dramatic impacts on final solution quality. Detailed routing seeks to refine global routes and typically does not alter the configuration of nets determined by global routing. Hence, if the global routing solution is poor, the quality of the detailed routing solution will likewise suffer.

To determine the net ordering, each net is given a numerical indicator of importance (priority), known as a *net weight*. High priority can be given to nets that are timing-critical, connect to numerous pins, or carry specific functions such as delivering clock signals. High-priority nets should avoid unnecessary detours, even at the cost of detouring other nets. Pin ordering is typically performed using either tree-based algorithms (Sec. 5.6.1) or geometric criteria based on pin locations.

Specializing routing into global and detailed stages is common for digital circuits. For analog circuits, multi-chip modules (MCMs), and printed circuit boards (PCBs), global routing is sometimes unnecessary due to the smaller number of nets involved, and only detailed routing is performed.

5.2 Terminology and Definitions

The following terms are relevant to global routing in general. Terms pertaining to specific algorithms and techniques will be introduced in their respective sections.

A *routing track* (*column*) is an available horizontal (vertical) wiring path. A signal net often uses a sequence of alternating horizontal tracks and vertical columns, where adjacent tracks and columns are connected by inter-layer vias.

A *routing region* is a region that contains routing tracks and/or columns.

A *uniform routing region* is formed by evenly spaced horizontal and vertical grid lines that induce a uniform grid over the chip area. This grid is sometimes referred to as a *ggrid* (global grid); it is composed of unit *gcells* (global cells). Grid lines are typically spaced seven to 40 routing tracks [5.18] apart to balance the complexities of the chip-scale global routing and gcell-scale detailed routing problems.

A *non-uniform routing region* is formed by horizontal and vertical boundaries that are aligned to external pin connections or macro-cell boundaries. This results in *channels* and *switchboxes* – routing regions that have differing sizes. During global routing, nets are assigned to these routing regions. During detailed routing, the nets within each routing region are assigned to specific wiring paths.

A *channel* is a rectangular routing region with pins on two opposite (usually the longer) sides and no pins on the other (usually the shorter) sides. There are two types of channels – *horizontal* and *vertical*.

A *horizontal channel* is a channel with pins on the top and bottom boundaries (Fig. 5.3).

A *vertical channel* is a channel with pins on the left and right boundaries (Fig. 5.4).

The pins of a net are connected to the routing channel by columns, which in turn are connected to other columns by tracks. In older, variable-die routing contexts, such as two-layer standard-cell routing, the channel height is flexible, i.e., its capacity can be adjusted to accommodate the necessary amount of wiring. However, due to the increased number of routing layers in modern designs, this traditional channel model has largely lost its relevance. Instead, *over-the-cell* (*OTC*) routing is used, as discussed further in Sec. 6.7.

The *channel capacity* represents the number of available routing tracks or columns. For single-layer routing, the capacity is the height h of the channel divided by the pitch d_{pitch}, where d_{pitch} is the minimum distance between two wiring paths in the relevant (vertical or horizontal) direction (Fig. 5.3). For multilayer routing, the capacity σ is the sum of the capacities of all layers.

$$\sigma(Layers) = \sum_{layer \in Layers} \left\lfloor \frac{h}{d_{pitch}(layer)} \right\rfloor$$

Here, *Layers* is the set of all layers, and $d_{pitch}(layer)$ is the routing pitch for *layer*.

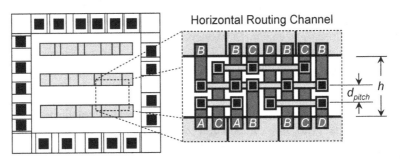

Fig. 5.3 Channel routing for a horizontal channel.

A *switchbox* (Fig. 5.4) is the intersection of horizontal and vertical channels. Due to fixed dimensions, switchbox routing exhibits less flexibility and is more difficult than channel routing. Note that since the entry points of wires are fixed, the problem is one of finding routes inside the switchbox.

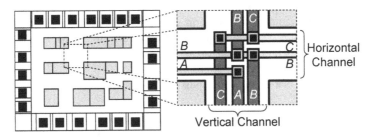

Fig. 5.4 Switchbox routing between horizontal and vertical channels of a macro-cell circuit.

A *2D switchbox* (Fig. 5.5) is a switchbox with terminals on four boundaries (top, bottom, left, and right). This model is primarily used with two-layer routing, where interlayer connections (vias) are relatively insignificant.

A *3D switchbox* (Fig. 5.5) is a switchbox with terminals on all six boundaries (top, bottom, left, right, up, and down), allowing paths to travel between routing layers. Four of the sides are the same as those of the 2D switchbox, one goes to the layer above, and one goes to the layer below.

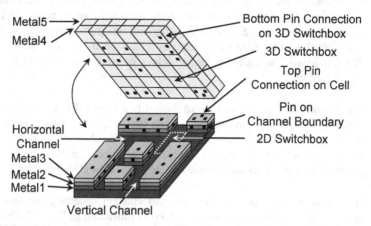

Fig. 5.5 2D and 3D switchboxes in a five-layer process. The cells are internally routed using layers up to Metal3. 2D switchboxes typically exist on layers Metal1, Metal2 and Metal3. Layers Metal4 and Metal5 are connected by a 3D switchbox.

A *T-junction* occurs when a vertical channel meets a horizontal channel, e.g., in a macro-cell placement. In the example of Fig. 5.6, both (1) the height of the vertical channel and (2) the locations of the fixed pin connections for the horizontal channel are determined only after the vertical channel has been routed. Therefore, the vertical channel must be routed before the horizontal channel.

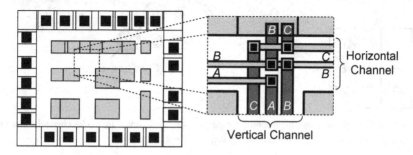

Fig. 5.6 Routing for a T-junction in a macro-cell placement.

5.3 Optimization Goals

Global routing seeks to (1) determine whether a given placement is routable, and (2) determine a coarse routing for all nets within available routing regions. In global routing, the *horizontal* and *vertical capacities* of a given routing region are, respectively, the maximum number of horizontal and vertical routes that can traverse that particular region. These upper bounds are determined by the specific semiconductor technology and its corresponding design rules, e.g., track pitches, as well as layout features. As the subsequent detailed routing will only have a fixed number of routing tracks, routing regions in global routing should not be oversubscribed. Optimization objectives addressed by routing include minimizing total wirelength and reducing signal delays on nets that are critical to the chip's overall timing.

Full-custom design. A full-custom design is a layout that is dominated by macro cells and wherein routing regions are non-uniform, often with different shapes and heights/widths.[2] In this context, two initial tasks, *channel definition* and *channel ordering*, must be performed. The channel definition problem seeks to divide the global routing area into appropriate routing channels and switchboxes.

To determine the types of channels and channel ordering, the layout region is represented by a floorplan tree (Sec. 3.3), as illustrated in Fig. 5.7. Divisions or cut lines in the layout, corresponding to routing channels, can be ordered according to a *bottom-up* (e.g., *post-order*) traversal of the internal nodes of the floorplan tree. This determines the routing sequence of the channels.

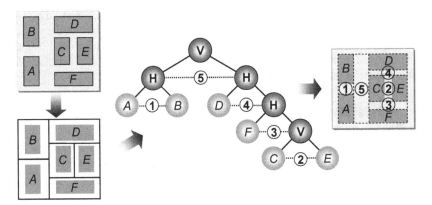

Fig. 5.7 Finding channels and their routing sequence using a floorplan tree. The channel ordering from 1 to 5 is determined from a bottom-up traversal of the internal nodes.

---

[2] The characteristics of a full-custom design differ among companies.

If (a part of) a floorplan is *non-slicing*, i.e., cannot be represented completely with only horizontal and vertical cuts, then at least one switchbox must be used, in addition to channels. Once all of the regions and their respective channels or switchboxes have been determined, the nets can be routed. Methods for routing nets include Steiner tree routing and minimum spanning tree routing (Sec. 5.6.1), as well as shortest-path routing using Dijkstra's algorithm (Sec. 5.6.3).

In full-custom, variable-die designs, the dimensions of routing regions are typically not fixed. Hence, the capacity of each routing region is not known *a priori*. In this context, the standard objective is to *minimize the total routed length* and/or *minimize the length of the longest timing path*. On the other hand, in the context of fixed-die placement, where the capacities of routing regions are constrained by hard upper bounds, routing optimization is often performed subject to a *routability* constraint.

Standard-cell design. In standard-cell designs, if the number of metal layers is limited, *feedthrough cells* must be used to route a net across multiple cell rows. Instantiating a feedthrough cell essentially reserves an empty vertical column for a net. Fig. 5.8 illustrates the use of feedthrough cells in the routing of a five-pin net. When feedthrough cells in consecutive rows are used during the routing of a given net, these cells must align so that their reserved tracks align.

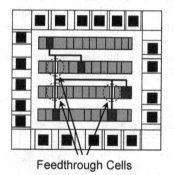

Feedthrough Cells

Fig. 5.8 The routing solution of the given net (cells in dark gray) with three feedthrough cells (white). Designs with more than three metal layers and over-the-cell routing typically do not use feedthrough cells.

If the cell rows and the netlist are fixed, then the number of unoccupied sites in the cell rows is fixed. Therefore, the number of possible feedthroughs is limited. Hence, standard-cell global routing seeks to (1) *ensure routability* of the design and (2) *find an uncongested solution* that minimizes total wirelength.

If a net is contained within a *single routing region*, then routing this net entails either channel or switchbox routing and can be achieved entirely within that region, except when significant detours are required due to routing congestion. However, if a net spans more than one routing region, then the global router must split the net into multiple *subnets*, and then assign the subnets to the routing regions. For multi-pin routing, rectilinear Steiner trees (Sec. 5.6.1) are commonly used (Fig. 5.9).

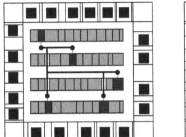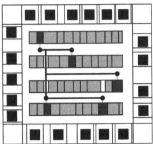

Fig. 5.9 Two different rectilinear Steiner tree solutions for routing of a five-pin net. The solution on the left has the least wirelength, while the solution on the right uses the fewest vertical segments.

The total height of a variable-die standard-cell design is the sum of all cell row heights (fixed quantity) plus all channel heights (variable quantities). Ignoring the small impact of feedthrough cells, minimizing the layout area is equivalent to minimizing the sum of channel heights. Thus, although the layout is primarily determined during placement, the global routing solution also influences the layout size. Shorter routes typically lead to a more compact layout.

Gate-array designs. In gate-array designs, the sizes of the cells and the sizes of the routing regions between the cells (routing capacities) are fixed. Since routability cannot be ensured by the insertion of additional routing resources, key tasks include determining the placement's *routability* and finding a *feasible routing solution*. Like other design styles, additional optimization objectives include *minimizing total routed wirelength* and *minimizing the length of the longest timing path*.

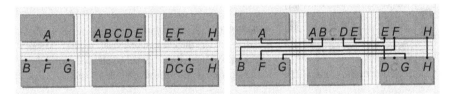

Fig. 5.10 A gate-array design with channel height = 4 (left). If all nets are routed with their shortest paths, the full netlist is unroutable. In one possible routing, net *C* is remains unrouted (right).

5.4 5.4 Representations of Routing Regions

To model the global routing problem, the routing regions (e.g., channels or switchboxes) are represented using efficient data structures. Typically, the routing context is captured using a graph, where nodes represent routing regions and edges represent adjoining regions.[3] For each prescribed connection, a router must

---

[3] Capacities are associated with both edges and nodes to represent available routing resources.

determine a path within the graph that connects the terminal pins. The path can only traverse nodes and edges of the graph that have sufficient remaining routing resources. The following three graph models are commonly used.

A *grid graph* is defined as *ggrid* = (V,E), where the nodes $v \in V$ represent the routing grid cells (*gcells*) and the edges represent connections of grid cell pairs (v_i,v_j) (Fig. 5.11). The global routing grid graph is two-dimensional, but must represent k routing layers. Hence, k distinct capacities must be maintained at each node of the grid graph. For example, for a two-layer routing grid ($k = 2$), a capacity pair (3,1) for a routing grid cell can represent three horizontal segments and one vertical segment still available. Other capacity representations are possible.

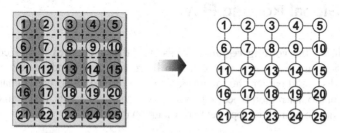

Fig. 5.11 A layout and its corresponding grid graph.

In a *channel connectivity graph* $G = (V,E)$, the nodes $v \in V$ represent *channels*, and the edges E represent adjacencies of the channels (Fig. 5.12). The capacity of each channel is represented in its respective graph node.

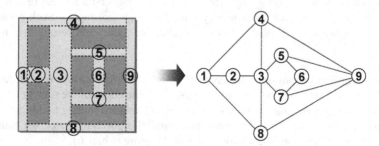

Fig. 5.12 Sample layout with its corresponding channel connectivity graph.

In a *switchbox connectivity* (*channel intersection*) *graph* $G = (V, E)$, the nodes $v \in V$ represent switchboxes. An edge exists between two nodes if the corresponding switchboxes are on opposite sides of the same channel (Fig. 5.13). In this graph model, the edges represent horizontal and vertical channels.

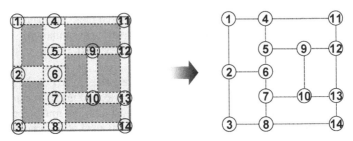

Fig. 5.13 Sample layout with its corresponding switchbox connectivity graph.

5.5 5.5 The Global Routing Flow

Step 1: defining the routing regions. In this step, the layout area is divided into routing regions. In some cases, nets can also be routed over standard cells (OTC routing). As discussed earlier, the routing regions are formed as 2D or 3D channels, switchboxes and other region types (Fig. 5.5). These routing regions, their capacities, and their connections are then represented by a graph (Sec. 5.4).

Step 2: mapping nets to the routing regions. In this step, each net of the design is tentatively assigned to one or several routing regions so as to connect all of its pins. The routing capacity of each routing region limits the number of nets traversing this region. Other factors, such as timing and congestion, also affect the path chosen for each net. For example, routing resources can be priced differently in different regions with available routing capacity – the more congested a routing region, the higher the cost for any net subsequently routed through that region. Such resource pricing encourages subsequent nets to seek alternate paths, and results in a more uniform routing density.

Step 3: assigning crosspoints. In this step, also known as *midway routing*, routes are assigned to fixed locations, or crosspoints, along the edges of the routing regions. Crosspoint assignment enables scaling of global and detailed routing to designs with millions of cells as well as distributed and parallel algorithms, since the routing regions can be handled independently in detailed routing (Chap. 6).

Finding an optimal crosspoint assignment requires knowledge of net connection dependencies and channel ordering. For instance, the crosspoints of a switchbox can be fixed only after all adjacent channels have been routed. Thus, the locations will depend on local connectivity through the channels and switchboxes. Note that some global routers do not perform a separate crosspoint assignment step. Instead, *implicit* crosspoint assignment is integrated with detailed routing.

5.6 Single-Net Routing

The following techniques for single-net routing are commonly used within larger full-chip routing tools.

▶ **5.6.1 Rectilinear Routing**

Multi-pin nets – nets with more than two pins – are often decomposed into two-pin subnets, followed by point-to-point routing of each subnet according to some ordering. Such net decomposition is performed at the beginning of global routing and can affect the quality of the final routing solution.

Rectilinear spanning tree. A rectilinear spanning tree connects all terminals (pins) using only *pin-to-pin connections* that are composed of vertical and horizontal segments. Pin-to-pin connections can meet only at a pin, i.e., "crossing" edges do not intersect, and no additional junctions (*Steiner points*) are allowed. If the total length of segments used to create the spanning tree is minimal, then the tree is a *rectilinear minimum spanning tree* (*RMST*). An RMST can be computed in $O(p^2)$ time, where p is the number of terminals in the net, using methods such as Prim's algorithm [5.19]. This algorithm builds an MST by starting with a single terminal and greedily adding least-cost edges to the partially-constructed tree until all terminals are connected. Advanced computational-geometric techniques reduce the runtime to $O(p \log p)$.

Rectilinear Steiner tree (RST). A rectilinear Steiner tree (RST) connects all p pin locations and possibly some additional locations (Steiner points). While any rectilinear spanning tree for a p-pin net is also a rectilinear Steiner tree, the addition of carefully-placed Steiner points often reduces the total net length.[4] An RST is a *rectilinear Steiner minimum tree* (*RSMT*) if the total length of net segments used to connect all p pins is minimal. For instance, in a uniform routing grid, let a *unit net segment* be an edge that connects two adjacent gcells; an RST is an RSMT if it has the minimum number of unit net segments.

The following facts are known about RSMTs.

– An RSMT for a p-pin net has between 0 and $p - 2$ (inclusive) Steiner points.
– The degree of any terminal pin is 1, 2, 3, or 4. The degree of a Steiner point is either 3 or 4.
– A RSMT is always enclosed in the *minimum bounding box* (*MBB*) of the net.
– The total edge length L_{RSMT} of the RSMT is at least half the perimeter of the minimum bounding box of the net: $L_{RSMT} \geq L_{MBB} / 2$.

---

[4] In Manhattan routing, the corner of an L-shape connection between two points is not considered a Steiner point.

Constructing RSMTs in the general case is NP-hard; in practice, heuristic methods are used. One fast technique, *FLUTE*, developed by *C. Chu* and *Y. Wong* [5.5], finds optimal RSMTs for up to nine pins and produces near-minimal RSTs, often within 1% of the minimum length, for larger nets. Although RSMTs are optimal in terms of wirelength, they are not always the best choice of net topology in practice.

Some heuristics do not deal well with obstacle avoidance or other requirements. Furthermore, the minimum wirelength objective may be secondary to control of timing or signal integrity on particular paths in the design. Therefore, nets may alternatively be connected using RMSTs, which can be computed in low-order polynomial time and are guaranteed to have no more than 1.5 times the net length of an RSMT. This worst-case ratio occurs, for example, with terminals at (1,0), (0,1), (-1,0) and (0,-1). The RSMT has a Steiner point at (0,0) and total net length = 4, while the RMST uses three length-2 edges with total net length = 6. Since RMSTs are computed optimally and quickly, many heuristic approaches, such as the one presented below, transform an initial RMST into a low-cost RSMT [5.10].

Example: RMSTs and RSMTs
Given: rectilinear minimum spanning tree (RMST) (right).
Task: transform the RMST into a heuristic RSMT.

Solution:
The transformation relies on the fact that for each two-pin connection, two different *L*-shapes may be formed. *L*-shapes that cause (more) overlap of net segments, and hence reduction of the total wirelength of the net, are preferred. Construct *L*-shapes between points p_1 and p_2.

Construct another set of *L*-shapes between p_2 and p_3. This introduces the Steiner point S_1.

No further wirelength reduction is possible. The final tree, which is an RSMT, has Steiner point S and is shown on the right.

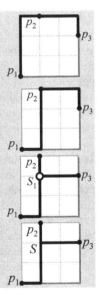

Hanan grid. The previous example showed that adding Steiner points to an RMST can significantly reduce the wirelength of the net. In 1966, *M. Hanan* [5.8] proved that in finding an RSMT, it suffices to consider only Steiner points located at the intersections of vertical and horizontal lines that pass through terminal pins. More formally, the *Hanan grid* (Fig. 5.14) consists of the lines $x = x_p$, $y = y_p$ that pass through each pin location (x_p, y_p). The Hanan grid contains at most p^2 candidate Steiner points, thereby greatly reducing the solution space for finding an optimal RSMT.

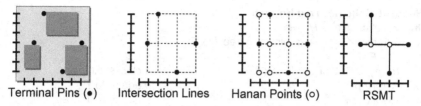

Terminal Pins (•)   Intersection Lines   Hanan Points (o)   RSMT

Fig. 5.14 Finding the Hanan grid and subsequently the Steiner points of an RSMT.

Defining the routing regions. For global routing with Steiner trees, the layout is usually divided into a coarse routing grid consisting of *gcells*, and represented by a graph (Sec. 5.4). The example in Fig. 5.15 assumes a standard-cell layout. Without explicitly defining channel height, the distances between the horizontal lines is used as an estimate. Based on this estimate, the distances between the vertical lines are chosen such that each grid cell has a 1:1 aspect ratio. All connection points within the grid cell are treated as if they are at the cell's midpoint.

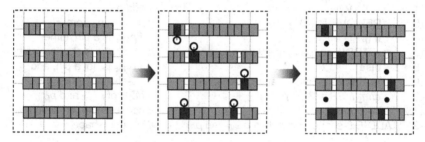

Fig. 5.15 In the global layout of standard cells (left), standard cells (dark gray) need to be connected at their pins (center). These pins are assigned to grid cells for Steiner tree construction (right).

Steiner tree construction heuristic. The following heuristic uses the Hanan grid to produce a close-to-minimal rectilinear Steiner tree. The heuristic greedily makes the shortest possible new connection consistent with a tree topology (recall Prim's minimum spanning tree algorithm), leaving as much flexibility as possible in the embedding of edges. For nets with up to four pins, this heuristic always produces RSMTs. Otherwise the resulting RSTs are not necessarily minimal.

The heuristic (pseudocode on the next page) works as follows. Let $P' = P$ be the set of pins to consider (line 1). Find the closest (in terms of rectilinear distance) pin pair (p_A, p_B) in P' (line 2), add them to the tree T, and remove them from P' (lines 3-6). If there are only two pins, then the shortest path connecting them is an L-shape (lines 7-8). Otherwise, construct the minimum bounding box (MBB) between p_A and p_B (line 10), and find the closest point pair (p_{MBB}, p_C) between MBB(p_A, p_B) and P', where p_{MBB} is the point located on MBB(p_A, p_B), and p_C is the pin in P' (line 12). If p_{MBB} is a pin, then add any L-shape to T (lines 16-17). Otherwise, add the L-shape that p_{MBB} lies on (lines 18-19). Construct the MBB of p_{MBB} and p_C (line 20), and repeat this process until P' is empty (lines 11-20). Finally, add the remaining (unconnected) pin to T (lines 21-22).

Sequential Steiner Tree Heuristic
Input: set of all pins P
Output: heuristic Steiner minimum tree $T(V,E)$

1.	$P' = P$	
2.	$(p_A,p_B) =$ CLOSEST_PAIR(P')	// closest pin pair
3.	ADD(V,p_A)	// add p_A to T
4.	ADD(V,p_B)	// add p_B to T
5.	REMOVE(P',p_A)	// remove p_A from P'
6.	REMOVE(P',p_B)	// remove p_B from P'
7.	**if** ($P' == \varnothing$)	// shortest path connecting
8.	ADD(E,L-shape connecting p_A and p_B)	// p_A and p_B is any L-shape
9.	**else**	
10.	$curr\_MBB =$ MBB(p_A,p_B)	// MBB of p_A and p_B
11.	**while** ($P' \neq \varnothing$)	
12.	$(p_{MBB},p_C) =$ CLOSEST_PAIR($curr\_MBB,P'$)	// closest point pair, one from
		// $curr\_MBB$, one from P'
13.	ADD(V,p_{MBB})	// add p_{MBB} to T
14.	ADD(V,p_C)	// add p_C to T
15.	REMOVE(P',p_C)	// remove p_C from P'
16.	**if** ($p_{MBB} \in P$)	// if p_{MBB} is a pin, either
17.	ADD(L-shape connecting p_{MBB} and p_C)	// L-shape is shortest path
18.	**else**	// if p_{MBB} is not a pin, add
19.	ADD(E,L-shape that includes p_{MBB})	// L-shape that p_{MBB} is on
20.	$curr\_MBB =$ MBB(p_{MBB},p_C)	// MBB of p_{MBB} and p_C
21.	ADD(V,p_C)	// connect T to remaining pin
22.	ADD(E,L-shape connecting p_{MBB} and p_C)	// with L-shape

---

Example: Sequential Steiner Tree Heuristic
Given: seven pins p_1-p_7 and their coordinates (right).

p_1 (0,6) p_2 (1,5) p_3 (4,7) p_4 (5,4) p_5 (6,2) p_6 (3,2) p_7 (1,0)

Task: construct a heuristic Steiner minimum tree using the sequential Steiner tree heuristic.

Solution:
$P' = \{p_1,p_2,p_3,p_4,p_5,p_6,p_7\}$
Pins p_1 and p_2 are the closest pair of pins. Remove them from P.
$P' = \{p_3,p_4,p_5,p_6,p_7\}$
Construct the MBB of p_1 and p_2, and find the closest point pair between MBB(p_1,p_2) and P', which are respectively $p_{MBB} = p_a$ and $p_C = p_3$. Since p_a is not a pin, select the L-shape that p_a lies on. Construct the MBB of p_a and p_3. Remove p_3 from P.

$P' = \{p_4,p_5,p_6,p_7\}$
Find the closest point pair between MBB(p_a,p_3) and P', which are respectively $p_{MBB} = p_b$ and $p_C = p_4$. Since p_b is not a pin, select the L-shape that p_b lies on. Construct the MBB of p_b and p_4. Remove p_4 from P'.

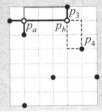

$P' = \{p_5, p_6, p_7\}$
Find the closest point pair between MBB(p_b, p_4) and P', which are respectively $p_{MBB} = p_4$ and $p_C = p_5$. Since p_4 is a pin, select either L-shape that connects p_b and p_4. In this example, the right-down L-shape is selected. Construct the MBB of p_4 and p_5. Remove p_5 from P.

$P' = \{p_6, p_7\}$
Find the closest point pair between MBB(p_4, p_5) and P', which are respectively $p_{MBB} = p_c$ and $p_C = p_6$. Since p_c is not a pin, select the L-shape that p_c lies on. Construct the MBB of p_c and p_6. Remove p_6 from P'.

$P' = \{p_7\}$
Find the closest point pair between MBB(p_c, p_6) and P', which are respectively $p_{MBB} = p_6$ and $p_C = p_7$. Since p_6 is a pin, select either L-shape that connects p_c and p_6. In this example, there is one shortest path between p_c and p_6. Construct the MBB of p_6 and p_7. Remove p_7 from P'.

$P' = \varnothing$
Since there is only one pin unconnected (p_7), select either L-shape that connects p_6 and p_7. In this example, the down-left L-shape is selected.

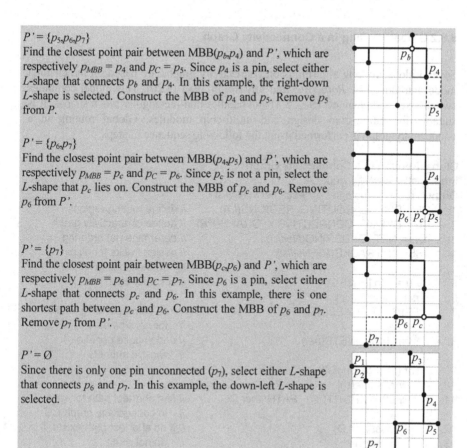

Mapping nets to routing regions. After finding an appropriate Steiner tree topology for the net, the segments are mapped to the physical layout. Each net segment is assigned to a specific grid cell (Fig. 5.16). When making these assignments, horizontal and vertical capacities for each grid cell are also considered.

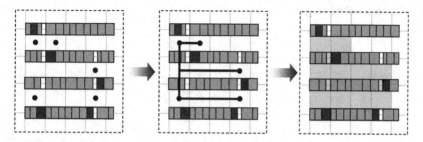

Fig. 5.16 Assignment of each segment of the Steiner tree generated for the net from Fig. 5.15 (left and center). The routing regions affected are shaded in light gray (right).

▶ 5.6.2 Global Routing in a Connectivity Graph

Several global routing algorithms on connectivity graphs are based on the channel model introduced by *Rothermel* and *Mlynski* in 1983 [5.20]. This model combines switchboxes and channels (Sec. 5.4) and handles non-rectangular block shapes. It is suitable for full-custom design and multi-chip modules. Global routing in a connectivity graph is performed using the following sequence of steps.

Global Routing in a Connectivity Graph
Input: netlist *Netlist*, layout *LA*
Output: routing topologies for each net in *Netlist*
1.	RR = DEFINE_ROUTING_REGIONS(*LA*)	// define routing regions
2.	CG = DEFINE_CONNECTIVITY_GRAPH(RR)	// define connectivity graph
3.	nets = NET_ORDERING(*Netlist*)	// determine net ordering
4.	ASSIGN_TRACKS(RR,*Netlist*)	// assign tracks for all pin
		// connections in *Netlist*
5.	**for** (i = 1 to |nets|)	// consider each net
6.	net = nets[*i*]	
7.	FREE_TRACKS(*net*)	// free corresponding tracks
		// for *net*'s pins
8.	snets = SUBNETS(*net*)	// decompose *net* into
		// two-pin subnets
9.	**for** (j = 1 **to** |snets|)	
10.	snet = snets[*j*]	
11.	spath = SHORTEST_PATH(*snet*,*CG*)	// find shortest path for *snet*
		// in connectivity graph *CG*
12.	**if** (spath == $\emptyset$)	// if no shortest path exists,
13.	**continue**	// do not route
14.	**else**	// otherwise, assign *snet* to
15.	ROUTE(*snet*,*spath*,*CG*)	// the nodes of *spath* and
		// update routing capacities

Defining the routing regions. The vertical and horizontal routing regions are formed by stretching the bounding box of cells in each direction until a cell or chip boundary is reached (Fig. 5.17).

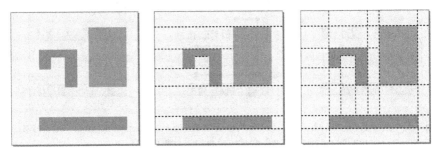

Fig. 5.17 Given a layout area with macro blocks (left), the horizontal (center) and vertical (right) macro-cell edges are extended to form the routing regions.

Defining the connectivity graph. In the connectivity graph representation (Fig. 5.18), the nodes represent the routing regions, and an edge between two nodes indicates that those routing regions are connected (continuity of the routing region). Each node also maintains the horizontal and vertical capacities for its routing region.

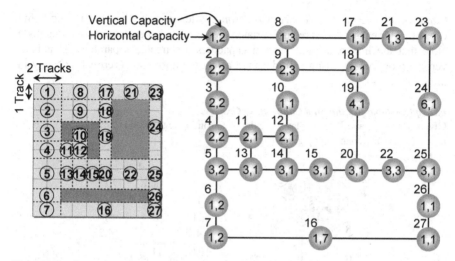

Fig. 5.18 Routing regions represented by a connectivity graph. Each node maintains the horizontal and vertical routing capacities for its corresponding routing region. The edges between the nodes show the connectivity between routing regions.

Determining the net order. The order in which nets are processed can be determined before or during routing. Nets can be prioritized according to criticality, number of pins, size of bounding box (larger means higher priority), or electrical properties. Some algorithms dynamically update priorities based on layout characteristics observed during the course of routing attempts.

Assigning tracks for all pin connections. For each pin *pin*, a horizontal track and a vertical track are reserved within *pin*'s routing region. This step is necessary to ensure that *pin* can be connected. It also provides two major advantages. First, since track assignment of pin connections is performed *before* global routing, if the pins are not accessible, then the placement of cells must be adjusted. Second, track reservation prevents nets that are routed first from blocking pin connections that are used later. This gives the router a more accurate congestion map and enables more intelligent detouring for future nets [5.1].

Global routing of all nets. Each net is processed separately in a pre-determined order. The following steps are applied to each net.

Net and/or subnet ordering: split the multi-pin net into two-pin subnets, then determine an appropriate order by sorting the pins of each subnet, e.g., in non-decreasing order with respect to the *x*-coordinates.

Track assignment in the connectivity graph: tracks are assigned using a maze-routing algorithm (Sec. 5.6.3), with the regions' remaining resources as weights. High congestion in a region will encourage paths to detour through regions with low congestion.

Capacity update in the connectivity graph: after a route has been found, the capacities for each region are appropriately decremented at each corresponding node. Note that the horizontal and vertical capacities are treated separately. That is, a vertical (horizontal) wiring path does not affect the horizontal (vertical) capacity in the same region.

Example: Global Routing in a Connectivity Graph
Given: (1) nets *A* and *B*, (2) the layout region with routing regions and obstacles (left), and (3) the corresponding connectivity graph (right).

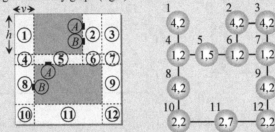

Task: route *A* and *B* using as few resources as possible.

Solution:
Route *A* before *B*. Note: track assignment of pin connections is omitted in this example.

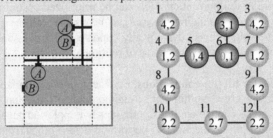

After *A* is routed, route *B*. Note that *B* is detoured since the horizontal capacities of nodes (regions) 5 and 6 are 0.

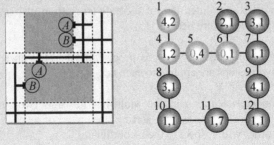

Example: Determining Routability
Given: (1) nets *A* and *B*, (2) the layout area with routing regions and obstacles (left), and (3) the corresponding connectivity graph (right).
Task: determine the routability of *A* and *B*.

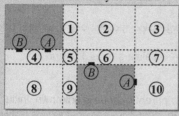 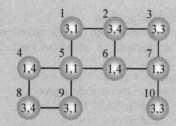

Solution:
Net *A* is first routed through nodes (regions) 4-5-6-7-10, which is a shortest path. After this assignment, the horizontal capacities of nodes 4-5-6-7 are exhausted, i.e., each horizontal capacity = 0.

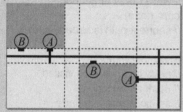 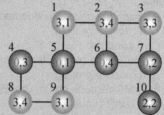

The shortest path for net *B* was previously through nodes 4-5-6, but this path would now make these nodes' horizontal capacities negative. A longer (but feasible) path exists through nodes 4-8-9-5-1-2-6. Thus, this particular placement is considered routable.

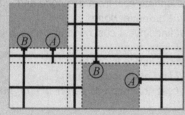 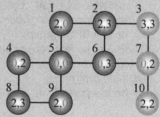

► **5.6.3 Finding Shortest Paths with Dijkstra's Algorithm**

Dijkstra's algorithm [5.6] finds shortest paths from a specified *source node* to all other nodes in a graph with non-negative edge weights. This is often referred to as *maze routing*. The algorithm is useful for finding a shortest path between two specific nodes in the routing graph – i.e., from a source to a particular target. In this context, the algorithm terminates when a shortest path to a target node is found.

Dijkstra's algorithm takes as input (1) a graph $G(V,E)$ with non-negative *edge weights W*, (2) a *source* (starting) node *s*, and (3) a *target* (ending) node *t*. The algorithm maintains three groups of nodes – (1) *unvisited*, (2) *considered*, and (3) *known*. Group 1 contains the nodes that have not yet been visited. Group 2 contains the nodes that have been visited but for which the shortest-path cost from the starting node *has not yet been found*. Group 3 contains the nodes that have been visited and for which the shortest path cost from the starting node *has been found*.

Dijkstra's Algorithm
Input: weighted graph $G(V,E)$ with edge weights W, source node s, target node t
Output: shortest path *path* from s to t

```
1.    group₁ = V                              // initialize groups 1, 2 and 3
2.    group₂ = group₃ = path = ∅
3.    foreach (node node ∈ group₁)
4.        parent[node] = UNKNOWN              // parent of node is unknown, initial
5.        cost[s][node] = ∞                   //   cost from s to any node is maximum
6.    cost[s][s] = 0                          //   except the s-s cost, which is 0
7.    curr_node = s                           // s is the starting node
8.    MOVE(s,group₁,group₃)                   // move s from Group 1 to Group 3
9.    while (curr_node != t)                  // while not at target node t
10.       foreach (neighboring node node of curr_node)
11.           if (node ∈ group₃)             // shortest path is already known
12.               continue
13.           trial_cost = cost[s][curr_node] + W[curr_node][node]
14.           if (node ∈ group₁)             // node has not been visited
15.               MOVE(node,group₁,group₂)    // mark as visited
16.               cost[s][node] = trial_cost  // set cost from s to node
17.               parent[node] = curr_node    // set parent of node
18.           else if (trial_cost < cost[s][node]) // node has been visited and
                                              //   new cost from s to node is lower
19.               cost[s][node] = trial_cost  // update cost from s to node and
20.               parent[node] = curr_node    //   parent of node
21.           curr_node = BEST(group₂)        // find lowest-cost node in Group 2
22.           MOVE(curr_node,group₂,group₃)   //   and move to Group 3
23.   while (curr_node != s)                  // backtrace from t to s
24.       ADD(path,curr_node)                 // add curr_node to path
25.       curr_node = parent[curr_node]       // set next node as parent of curr_node
```

First, all nodes are moved into Group 1 (line 1), while Groups 2 and 3 are empty (line 2). For each node *node* in Group 1, the cost of reaching *node* from the source node s is initialized to be infinite (∞), i.e., unknown, with the exception of s itself, which has cost 0. The parents of each node *node* in Group 1 are initially unknown since no node has yet been visited (lines 3-6). The source node s is set as the current node *curr_node* (line 7), which is then moved into Group 3 (line 8). While the target node t has not been reached (line 9), the algorithm computes the cost of reaching each neighboring node *node* of *curr_node* from s. The cost of reaching *node* from s is defined as the cost of reaching *curr_node* from s plus the cost of the edge from *curr_node* to *node* (lines 10-13). The former value is maintained by the algorithm, while the latter value is defined by edge weights.

If the neighboring node *node*'s shortest-path cost has already been computed, then move onto another neighboring node (lines 11-12). If *node* has not been visited, i.e., is in Group 1, the cost of reaching *node* from *s* is recorded and the parent of *node* is set to *curr_node*. (lines 14-17). Otherwise, the current cost from *s* to *node* is compared to the new cost *trial_cost*. If *trial_cost* is lower than the current cost from *s* to *node*, then the cost and parent of *node* are updated (lines 18-20). After all neighbors have been considered, the algorithm selects the best or lowest-cost node and sets it to be the new *curr_node* (line 21-22).

Once *t* is found, the algorithm finds the shortest path by backtracing. Starting with *t*, the algorithm visits the parent of *t*, visits the parent of that node, and so on, until *s* is reached (lines 23-25).

After the first iteration, Group 3 contains a non-source node that is closest (with respect to cost) to *s*. After the second iteration, Group 3 contains the node that has second-lowest shortest-path cost from *s*, and so on. Whenever a new node *node* is added to Group 3, the shortest path from *S* to *node* must pass only through nodes that are already in Group 3. This *invariant* guarantees a shortest-path cost.

The key to the efficiency of Dijkstra's algorithm is the small number of cost updates. During each iteration, only the neighbors of the node that was most recently added to Group 3 (*curr_node*) need to be considered. Nodes adjacent to *curr_node* are added to Group 2 if they are not already in Group 2. Then, the node in Group 2 with minimum shortest-path cost is moved to Group 3. Once the target node, i.e., the stopping point, has been added to Group 3, the algorithm traces back from the target node to the starting node to find the optimal (minimum-cost) path.

Dijkstra's algorithm not only guarantees a shortest (optimal) path based on the given non-negative edge weights, but can also optimize a variety of objectives, as long as they can be represented by edge weights. Examples of these objectives include geometric distance, electrical properties, routing congestion, and wire densities.

Example: Dijkstra's Algorithm
Given: graph with nine nodes *a-i* and edge weights (w_1, w_2) (right).
Task: find the shortest path using Dijkstra's algorithm from source *s* (node *a*) to target *t* (node *h*), where the path cost from node *A* to node *B* is

$$cost[A][B] = \sum w_1(A,B) + \sum w_2(A,B)$$

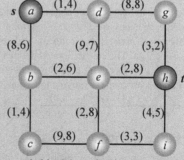

Solution:
Only Groups 2 and 3 are listed, as Group 1 is directly encoded in the graph. The parent[*node*] of each node *node* is its predecessor in the shortest path from *s*. Each *node*, with the exception of *s*, has the following format.

<parent of node> [node name] $(\sum w_1(s,node), \sum w_2(s,node))$

For example, <*a*> [*b*] (8,6) means that node *a* is the parent of node *b*, and has costs (8,6).

Iteration 1: *curr_node* = *a*
Add the starting node *s* = *a* to Group 3.

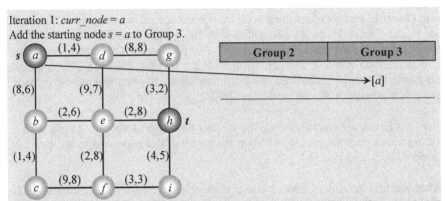

Find the cumulative path costs through the current node *a* to all its neighboring nodes (*b* and *d*). Add each neighboring node to Group 2, keeping track of its costs and parent.
b: *cost*[*s*][*b*] = 8 + 6 = 14, *parent*[*b*] = *a*
d: *cost*[*s*][*d*] = 1 + 4 = 5, *parent*[*d*] = *a*
Between *b* and *d*, *d* has the lower cost. Therefore, it is selected as the node to be moved from Group 2 to Group 3.

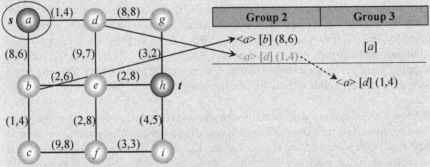

Iteration 2: *curr_node* = *d*
Compute the costs of all nodes that are adjacent to *d* but *not* in Group 3. If the neighboring node *node* is unvisited (*node* is in Group 1), then add it to Group 2 and set its costs and parent. Otherwise (*node* is in Group 2), update its costs and parent if its current cost is less than existing cost. The cost to *node* is defined as the minimum of (1) the existing path cost and (2) the path cost from *s* to *d* plus the edge cost from *d* to *node*. From Group 2, select the node with the least cost and move it to Group 3.

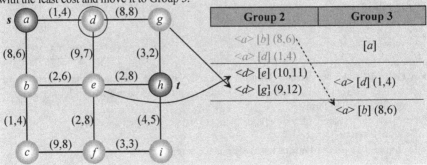

Iterations 3-6: Similar to Iteration 2. In Iteration 3, the entry *<b>* [e] (10,12) is rejected because the previous entry *<d>* [e] (10,11) is less. The end result is illustrated below.

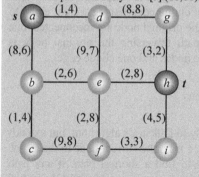

Group 2	Group 3
<a> [b] (8,6)	[a]
<a> [d] (1,4)	
<d> [e] (10,11)	*<a>* [d] (1,4)
<d> [g] (9,12)	
<b> [c] (9,10)	*<a>* [b] (8,6)
<b> [e] (10,12)	
<c> [f] (18,18)	*<b>* [c] (9,10)
<e> [f] (12,19)	*<d>* [e] (10,11)
<e> [h] (12,19)	
<g> [h] (12,14)	*<d>* [g] (9,12)
	<g> [h] (12,14)

Retrace from *t* to *s*.

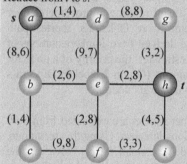

Group 2	Group 3
<a> [b] (8,6)	[a]
<a> [d] (1,4)	
<d> [e] (10,11)	*<a>* [d] (1,4)
<d> [g] (9,12)	
<b> [c] (9,10)	*<a>* [b] (8,6)
<b> [e] (10,12)	
<c> [f] (18,18)	*<b>* [c] (9,10)
<e> [f] (12,19)	*<d>* [e] (10,11)
<e> [h] (12,19)	
<g> [h] (12,14)	*<d>* [g] (9,12)
	<g> [h] (12,14)

Result:
Optimal path *a-d-g-h* from *s* = *a* to *t* = *h* with accumulated cost[a] (12, 14).

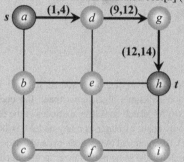

Dijkstra's algorithm can be applied to graphs where nodes also have a cost of traversal. To enable this, each capacitated node of degree d is replaced by a *d-clique* (a complete subgraph with d nodes), with each clique edge having weight equal to the original node cost. The d edges incident to the original node are reconnected, one per each of the d clique nodes. Thus, any path traversing the original node is transformed into a path that traverses two clique nodes and one clique edge.

▶ 5.6.4 Finding Shortest Paths with A* Search

The *A* search* algorithm [5.9] operates similarly to Dijkstra's algorithm, but extends the cost function to include an estimated distance from the current node to the target. Like Dijkstra's algorithm, A* also guarantees to find a shortest path, if any path exists, as long as the estimated distance to the target never exceeds the actual distance (i.e., an *admissibility* or *lower bound* criterion for the distance function). As illustrated in Fig. 5.19, A* search expands only the most promising nodes; its best-first search strategy eliminates a large portion of the solution space that is processed by Dijkstra's algorithm. A distance estimate that is a tight (accurate) lower bound on actual distance can lead to significant runtime improvements over *breadth-first search* (*BFS*) and its variants, and over Dijkstra's shortest-path algorithm. Implementations of A* search may be derived from implementations of Dijkstra's algorithm by adding distance-to-target estimates: the priority of a node for expansion in A* search is based on the lowest sum of (Group 2 label in Dijkstra's algorithm) + (distance estimate, including vias, to the target node).

A common variant is *bidirectional A* search*, where nodes are expanded from both the source and target until the two expansion regions intersect. Using this technique, the number of nodes considered can be reduced by a small factor. However, the overhead for bookkeeping, e.g., keeping track of the order in which the nodes have been visited, and the complexity of efficient implementation are challenging. In practice, this can negate the potential advantages of bidirectional search.

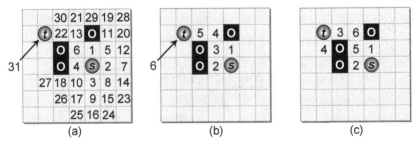

Fig. 5.19 An instance of shortest-path routing with source s, target t, and obstacles (solid black squares with 'O'). (a) BFS and Dijkstra's algorithm expand the nodes outward until t is found, exploring a total of 31 nodes. (b) A* search considers six nodes in the direction of t. When A* search is finished, it has considered only about a quarter as many nodes (six out of 31, in this example) that Dijkstra's algorithm does. (c) Bidirectional A* search expands nodes from both s and t. In this example, bidirectional A* search has the same performance as unidirectional A* search.

5.7 Full-Netlist Routing

In order to successfully route multiple nets, global routers must properly match nets with routing resources, without oversubscribing resources in any part of the chip. All signal nets are either routed *simultaneously*, e.g., using (integer) linear programming (Sec. 5.7.1), or *sequentially*, e.g., one net at a time (Sec. 5.6). When certain nets cause resource contention or overflow for routing edges, sequential routing requires multiple iterations. These iterations are performed by *ripping up* the nets that cause violations (Sec. 5.7.2) and *rerouting* them with fewer violations. The iterations continue until all nets are routed without violating capacities of routing-grid edges or until a timeout is exceeded.

▶ ### 5.7.1 Routing by Integer Linear Programming

A *linear program* (*LP*) consists of a set of *constraints* and an optional *objective function*. This function is maximized or minimized subject to these constraints. Both the constraints and the objective function must be linear. In particular, the constraints form a system of linear equations and inequalities. An *integer linear program* (*ILP*) is a linear program where every variable can only assume integer values. ILPs where all variables are binary are called *0-1 ILPs*. (Integer) Linear programs can be solved using a variety of available software tools such as *GLPK* [5.7], *CPLEX* [5.13], and *MOSEK* [5.17]. There are several ways to formulate the global routing problem as an ILP, one of which is presented below.

The ILP takes three inputs – (1) an $W \times H$ routing grid G, (2) routing edge capacities, and (3) the netlist *Netlist*. For exploitation purposes, a horizontal edge is considered to run left to right – $G(i,j) \sim G(i+1,j)$ – and a vertical edge is considered to run bottom to top – $G(i,j) \sim G(i,j+1)$.

The ILP uses two sets of variables. The first set contains k Boolean variables x_{net_1}, $x_{net_2}, \ldots, x_{net_k}$, each of which serves as an indicator for one of k specific paths or route options, for each net *net* $\in$ *Netlist*. If $x_{net_k} = 1$, (respectively, $= 0$), then the route option net_k is used (respectively, not used). The second set contains k real variables $w_{net_1}, w_{net_2}, \ldots, w_{net_k}$, each of which represents a net weight for a specific route option for *net* $\in$ *Netlist*. This net weight reflects the desirability of each route option for *net* (a larger w_{net_m} means that the route option net_m is more desirable – e.g., has fewer bends). With |*Netlist*| nets, and k available routes for each net *net* $\in$ *Netlist*, the total number of variables in each set is $k \cdot$ |*Netlist*|.

Next, the ILP formulation relies on two types of constraints. First, each net must select a single route (mutual exclusion). Second, to prevent overflows, the number of routes assigned to each edge (total usage) cannot exceed its capacity. The ILP maximizes the total number of nets routed, but may leave some nets unrouted. That is, if a selected route causes overflow in the existing solution, then the route will not be chosen. If all routes for a particular net cause overflow, then no routes will be chosen and thus the net will not be routed.

Integer Linear Programming (ILP) Global Routing Formulation

Inputs:

W,H	:	width W and height H of routing grid G
$G(i,j)$	:	grid cell at location (i,j) in routing grid G
$\sigma(G(i,j){\sim}G(i+1,j))$	:	capacity of horizontal edge $G(i,j) \sim G(i+1,j)$
$\sigma(G(i,j){\sim}G(i,j+1))$	:	capacity of vertical edge $G(i,j) \sim G(i,j+1)$
Netlist	:	netlist

Variables:

$x_{net_1}, \dots, x_{net_k}$	:	k Boolean path variables for each net *net* $\in$ *Netlist*
$w_{net_1}, \dots, w_{net_k}$	:	k net weights, one for each path of net *net* $\in$ *Netlist*

Maximize:

$$\sum_{net \in Netlist} w_{net_1} \cdot x_{net_1} + \dots + w_{net_k} \cdot x_{net_k}$$

Subject to:

Variable Ranges:

$$x_{net_1}, \dots, x_{net_k} \in [0,1] \qquad \forall net \in Netlist$$

Net Constraints:

$$x_{net_1} + \dots + x_{net_k} \leq 1 \qquad \forall net \in Netlist$$

Capacity Constraints:

$$\sum_{net \in Netlist} x_{net_1} + \dots + x_{net_k} \leq \sigma(G(i,j) \sim G(i,j+1)) \qquad \begin{array}{l}\forall net_k \text{ that use } G(i,j) \sim G(i,j+1), \\ 0 \leq i < W, 0 \leq j < H-1\end{array}$$

$$\sum_{net \in Netlist} x_{net_1} + \dots + x_{net_k} \leq \sigma(G(i,j) \sim G(i+1,j)) \qquad \begin{array}{l}\forall net_k \text{ that use } G(i,j) \sim G(i+1,j), \\ 0 \leq i < W-1, 0 \leq j < H\end{array}$$

In practice, most pin-to-pin connections are routed using *L*-shapes or *straight* wires (connections without bends). In this formulation, straight connections can be routed using a straight path or a *U*-shape; non-straight connections can use both *L*-shapes. For unrouted nets, other topologies can be found using *maze routing* (Sec. 5.6.3).

ILP-based global routers include *Sidewinder* [5.12] and *BoxRouter 1.0* [5.4]. Both decompose multi-pin nets into two-pin nets using *FLUTE* [5.5], and the route of each net is selected from two alternatives or left unselected. If neither of the two routes available for a net is chosen, Sidewinder performs maze routing to find an alternate route and replaces one of the unused routes in the ILP formulation. On the other hand, nets that were successfully routed and do not interfere with unrouted nets can be removed from the ILP formulation. Thus, Sidewinder solves multiple ILPs until no further improvement is observed. In contrast, BoxRouter 1.0 post-processes the results of its ILP using maze-routing techniques.

Example: Global Routing Using Integer Linear Programming
Given: (1) nets *A-C*, (2) an $W = 5 \times H = 4$ routing grid G, (3) $\sigma(e) = 1$
for all $e \in G$, and (4) *L*-shapes have weight 1.00 and *Z*-shapes have
weight 0.99. The lower-left corner is (0,0).
Task: write the ILP to route the nets in the graph to the right.

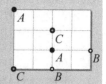

Solution:
For net *A*, the possible routes are two *L*-shapes (A_1, A_2) and two *Z*-shapes (A_3, A_4).

Net Constraints:
$x_{A_1} + x_{A_2} + x_{A_3} + x_{A_4} \leq 1$
Variable Constraints:
$0 \leq x_{A_1} \leq 1, 0 \leq x_{A_2} \leq 1,$
$0 \leq x_{A_3} \leq 1, 0 \leq x_{A_4} \leq 1$

For net *B*, the possible routes are two *L*-shapes (B_1, B_2) and one *Z*-shape (B_3).

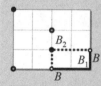
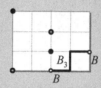

Net Constraints:
$x_{B_1} + x_{B_2} + x_{B_3} \leq 1$
Variable Constraints:
$0 \leq x_{B_1} \leq 1, 0 \leq x_{B_2} \leq 1,$
$0 \leq x_{B_3} \leq 1$

For net *C*, the possible routes are two *L*-shapes (C_1, C_2) and two *Z*-shapes (C_3, C_4).

Net Constraints:
$x_{C_1} + x_{C_2} + x_{C_3} + x_{C_4} \leq 1$
Variable Constraints:
$0 \leq x_{C_1} \leq 1, 0 \leq x_{C_2} \leq 1,$
$0 \leq x_{C_3} \leq 1, 0 \leq x_{C_4} \leq 1$

Each edge must satisfy capacity constraints. Only non-trivial constraints are shown.

Horizontal Edge Capacity Constraints:

$G(0,0) \sim G(1,0)$	:	$x_{C_1} + x_{C_3}$	$\leq$	$\sigma(G(0,0) \sim G(1,0)) = 1$
$G(1,0) \sim G(2,0)$	:	x_{C_1}	$\leq$	$\sigma(G(1,0) \sim G(2,0)) = 1$
$G(2,0) \sim G(3,0)$	:	$x_{B_1} + x_{B_3}$	$\leq$	$\sigma(G(2,0) \sim G(3,0)) = 1$
$G(3,0) \sim G(4,0)$	:	x_{B_1}	$\leq$	$\sigma(G(3,0) \sim G(4,0)) = 1$
$G(0,1) \sim G(1,1)$	:	$x_{A_2} + x_{C_4}$	$\leq$	$\sigma(G(0,1) \sim G(1,1)) = 1$
$G(1,1) \sim G(2,1)$	:	$x_{A_2} + x_{A_3} + x_{C_4}$	$\leq$	$\sigma(G(1,1) \sim G(2,1)) = 1$
$G(2,1) \sim G(3,1)$	:	x_{B_2}	$\leq$	$\sigma(G(2,1) \sim G(3,1)) = 1$
$G(3,1) \sim G(4,1)$	:	$x_{B_2} + x_{B_3}$	$\leq$	$\sigma(G(3,1) \sim G(4,1)) = 1$
$G(0,2) \sim G(1,2)$	:	$x_{A_4} + x_{C_2}$	$\leq$	$\sigma(G(0,2) \sim G(1,2)) = 1$
$G(1,2) \sim G(2,2)$	:	$x_{A_4} + x_{C_2} + x_{C_3}$	$\leq$	$\sigma(G(1,2) \sim G(2,2)) = 1$
$G(0,3) \sim G(1,3)$	:	$x_{A_1} + x_{A_3}$	$\leq$	$\sigma(G(0,3) \sim G(1,3)) = 1$
$G(1,3) \sim G(2,3)$	:	x_{A_1}	$\leq$	$\sigma(G(1,3) \sim G(2,3)) = 1$

Vertical Edge Capacity Constraints:

$$
\begin{array}{llll}
G(0,0) \sim G(0,1) & : & x_{C_2} + x_{C_4} & \leq & \sigma(G(0,0) \sim G(0,1)) = 1 \\
G(1,0) \sim G(1,1) & : & x_{C_3} & \leq & \sigma(G(1,0) \sim G(1,1)) = 1 \\
G(2,0) \sim G(2,1) & : & x_{B_2} + x_{C_1} & \leq & \sigma(G(2,0) \sim G(2,1)) = 1 \\
G(3,0) \sim G(3,1) & : & x_{B_3} & \leq & \sigma(G(3,0) \sim G(3,1)) = 1 \\
G(4,0) \sim G(4,1) & : & x_{B_1} & \leq & \sigma(G(4,0) \sim G(4,1)) = 1 \\
G(0,1) \sim G(0,2) & : & x_{A_2} + x_{C_2} & \leq & \sigma(G(0,1) \sim G(0,2)) = 1 \\
G(1,1) \sim G(1,2) & : & x_{A_3} + x_{C_3} & \leq & \sigma(G(1,1) \sim G(1,2)) = 1 \\
G(2,1) \sim G(2,2) & : & x_{A_1} + x_{A_4} + x_{C_1} + x_{C_4} & \leq & \sigma(G(2,1) \sim G(2,2)) = 1 \\
G(0,2) \sim G(0,3) & : & x_{A_2} + x_{A_4} & \leq & \sigma(G(0,2) \sim G(0,3)) = 1 \\
G(1,2) \sim G(1,3) & : & x_{A_3} & \leq & \sigma(G(1,2) \sim G(1,3)) = 1 \\
G(2,2) \sim G(2,3) & : & x_{A_1} & \leq & \sigma(G(2,2) \sim G(2,3)) = 1 \\
\end{array}
$$

Objective Function:

Maximize

$$
\begin{aligned}
& x_{A_1} + x_{A_2} + 0.99 \cdot x_{A_3} + 0.99 \cdot x_{A_4} \\
+ \; & x_{B_1} + x_{B_2} + 0.99 \cdot x_{B_3} \\
+ \; & x_{C_1} + x_{C_2} + 0.99 \cdot x_{C_3} + 0.99 \cdot x_{C_4}
\end{aligned}
$$

► **5.7.2 Rip-Up and Reroute (RRR)**

Modern ILP solvers help advanced ILP-based global routers to successfully complete hundreds of thousands of routes within hours [5.4][5.12]. However, commercial EDA tools require greater scalability and lower runtimes. These performance requirements are typically satisfied using the *rip-up and reroute* (*RRR*) framework, which focuses on problematic nets. If a net cannot be routed, this is often due to physical obstacles or other routed nets being in the way. The key idea is to allow *temporary* violations, so that all nets are routed, but then iteratively remove some nets (rip-up), and route them differently (reroute) so as to decrease the number of violations. In contrast, *push-and-shove* strategies [5.16] move currently routed nets to new locations (without rip-up) to relieve wire congestion or to allow previously unroutable nets to become routable.

An intuitive, greedy approach to routing would route nets sequentially and insist on violation-free routes where such routes are possible, even at the cost of large detours. On the other hand, the RRR framework allows nets to (temporarily) route through over-capacity regions.[5] This helps decide which nets should detour, rather than detouring the net routed most recently. In the example of Fig. 5.20(a), assume that the nets are routed in an order based on the size of the net's aspect ratio and MBB (*A-B-C-D*). If each net is routed without violations (Fig. 5.20(b)), then net *D* is forced to detour heavily. However, if nets are allowed to route with violations, then some nets are ripped up and rerouted, enabling *D* to use fewer routing segments (Fig. 5.20(c)).

---

[5] Allowing temporary violations is a common tactic for routing large-scale modern (ASIC) designs, while routing nets without violations is common for PCBs.

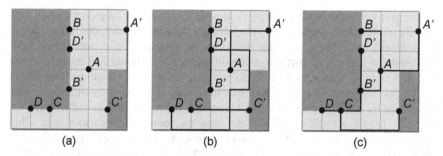

(a)　　　　　　　　　(b)　　　　　　　　　(c)

Fig. 5.20 Routing without violations versus rip-up and reroute. Let the net ordering be *A-B-C-D*. (a) The routing instance with nets *A-D*. (b) If all nets are routed without allowing violations, net *D* is forced to detour heavily; the resulting total wirelength is 21. (c) If routes are routed with allowing violations, all nets with violation can be ripped up and rerouted, resulting in lower wirelength. In this example, net *A* is rerouted with a shortest-path configuration, nets *B* and *C* are slightly detoured, and net *D* remains the same with its shortest-path configuration. The resulting wirelength is 19.

Traditional RRR strategies depend on (1) ripping up and rerouting all nets in violation, and (2) selecting an effective net ordering. That is, the order in which nets are routed greatly affects the quality of the final solution. For example, *Kuh* and *Ohtsuki* in 1990 defined quantifiable probabilities of RRR success (*success rates*) for each net in violation, and only ripped-up and rerouted the most promising nets [5.14]. However, these success rates are computed whenever a net could not be routed without violation, thereby incurring runtime penalties, especially for large-scale designs.

Global Routing Framework (With a Focus on Rip-Up and Reroute)
Input: unrouted nets *Netlist*, routing grid *G*
Output: routed nets *Netlist*

```
1.    v_nets = Ø                              // ordered list of violating nets
2.    foreach (net net ∈ Netlist)             // initial routing
3.       ROUTE(net,G)                          // route allowing violations
4.       if (HAS_VIOLATION(net,G))             // if net has violation,
5.          ADD_TO_BACK(v_nets,net)            //    add net to ordered list
                                               // start RRR framework
6.    while (v_nets ≠ Ø || !OUT())             // if nets still have violations or
                                               //    stopping condition is not met
7.       v_nets = REORDER(v_nets)              // optionally change net ordering
8.       for (i = 1 to i = |v_nets|)
9.          net = FIRST_ELEMENT(v_nets)        // process first element
10.         if (HAS_VIOLATION(net,G))          // net still has violation
11.            RIP_UP(net,G)                    // rip up net
12.            ROUTE(net,G)                     // reroute
13.            if (HAS_VIOLATION(net,G))        // if still has violation, add to
14.               ADD_TO_BACK(v_nets,net)       //    list of violating nets
15.            REMOVE_FIRST_ELEMENT(v_nets)     // remove first element
```

To improve computational scalability, a modern global router keeps track of all nets that are routed with violations – the nets go through at least one edge that is over-capacity. All these nets are added to an ordered list *v_nets* (lines 1-5). Optionally, *v_nets* can be sorted to suit a different ordering (line 7). For each net *net* in *v_nets* (line 8), the router first checks whether *net* still has violations (line 10). If *net* has no violations, i.e., some other nets have been rerouted away from congested edges used by *net*, then *net* is skipped. Otherwise, the router rips up and reroutes *net* (lines 11-12). If *net* still has violations, then the router adds *net* to *v_nets*. This process continues until all nets have been processed or a stopping condition is reached (lines 6-15). Variants of this framework include (1) ripping up all violating nets at once, and then rerouting nets one by one, and (2) checking for violations after rerouting all nets.

Notice that in this RRR framework, not all nets are necessarily ripped up. To further reduce runtime, some violating nets can be selectively chosen (temporarily) *not* to be ripped up. This typically causes wirelength to increase by a small amount, but reduces runtime by a large amount [5.11]. In the context of *negotiated congestion routing* (Sec. 5.8.2), nets are ripped-up and rerouted to also build up appropriate *history costs* on congested edges. Maintaining these history costs improves the success of rip-up and reroute and decreases the significance of ordering.

5.8   5.8 Modern Global Routing

As chip complexity grows, routers must limit both routed interconnect length and the number of vias, as this greatly affects the chip's performance, dynamic power consumption, and yield. Violation-free global routing solutions facilitate smooth transitions to *design for manufacturability* (*DFM*) optimizations. Completing global routing without violations allows the physical design process to move on to detailed routing and ensuing steps of the flow. However, if a placed design is inevitably unroutable or if a routed design exhibits violations, then a secondary step must isolate problematic regions. In cases where numerous violations are found, repair is commonly performed by repeating global or detailed placement and injecting whitespace into congested regions.

Several notable global routers have been developed for the *ISPD* 2007 and 2008 Global Routing Contests [5.18]. In 2007, *FGR* [5.21], *MaizeRouter* [5.16], and *BoxRouter* [5.4] claimed the top three places. In 2008, *NTHU-Route 2.0* [5.2] and *NTUgr* [5.3], which focused on better solution quality, and *FastRoute 3.0* [5.23], which focused on runtime, took the top three places.[6] Fig. 5.21 shows the general flow for several global routers, where each router uses a unique set of optimizations targeting a particular tradeoff between runtime and solution quality.

---

[6] *FastRoute 4.0* [5.22] was released shortly after the contest, with both solution quality and runtime improvements compared to FastRoute 3.0.

Given a *global routing instance* – a netlist and a routing grid with capacities – a global router first splits nets with three or more pins into two-pin subnets. It then produces an initial routing solution on a two-dimensional grid. If the design has no violations, the global router performs layer assignment – mapping 2D routes onto a 3D grid. Otherwise, nets that cause violations are ripped up and rerouted. This iterative process continues until the design is violation-free or a stopping condition (e.g., CPU limit) is reached. After rip-up and reroute, some routers perform an optional clean-up pass to further minimize wirelength. Other global routers directly route on the 3D grid; this method tends to improve wirelength, but is slow and may fail to complete routing. More information on the global routing flow, optimizations and implementation can be found in [5.11].

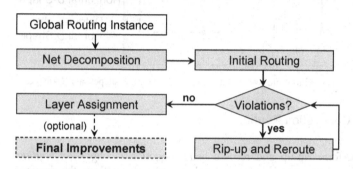

Fig. 5.21 Standard global routing flow.

Individual nets are often routed by constructing point-to-point connections using *maze routing* (Sec. 5.6.3) and *pattern routing* (Sec. 5.8.1) [5.22]. A popular method to control each net's cost is *negotiated congestion routing* (*NCR*) (Sec. 5.8.2) [5.15].

▶ **5.8.1 Pattern Routing**

Given a set of two-pin (sub)nets, a global router must find paths for each net while respecting capacity constraints. Most nets are routed with short paths to minimize wirelength. *Maze-routing* techniques such as Dijkstra's algorithm and A* search can be used to guarantee a shortest path between two points. However, these techniques can be unnecessarily slow, especially when the generated topologies are composed of edges (point-to-point connections) that are routed using very few vias, such as an *L*-shape. In practice, many nets' routes are not only short but also have few bends. Therefore, few nets require a maze router.

To improve runtime, *pattern routing* searches through a small number of route patterns. It often finds paths that cannot be improved, rendering maze routing unnecessary. Given an $m \times n$ bounding box where $n = k \cdot m$ and k is a constant, pattern routing takes $O(n)$ time, while maze routing requires $O(n^2 \log n)$ time. Topologies commonly used in pattern routing include *L*-shapes, *Z*-shapes, and *U*-shapes (Fig. 5.22).

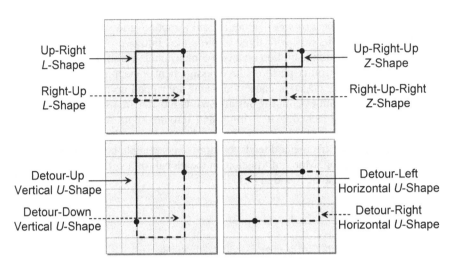

Fig. 5.22 Common patterns used to route two-pin nets such as *L*-shapes, *Z*-shapes, and *U*-shapes.

▶ **5.8.2 Negotiated Congestion Routing**

Modern routers perform rip-up and reroute using *negotiated congestion routing* (*NCR*), where each edge e is assigned a cost value $cost(e)$ that reflects the demand for edge e. A segment from net *net* that is routed through e pays a cost of $cost(e)$. The total cost of *net* is the sum of $cost(e)$ values taken over all edges used by *net*.

$$cost(net) = \sum_{e \in net} cost(e)$$

A higher $cost(e)$ value discourages nets from using e and implicitly encourages nets to seek out other, less used edges. Iterative routing approaches use methods such as Dijkstra's algorithm or A* search to find routes with minimum cost while respecting edge capacities. That is, during the current iteration, all nets are routed based on the current edge costs. If any nets cause violations, i.e., some edges e are congested, then (1) the nets are ripped up, (2) the costs of edges that these nets cross are updated to reflect their congestion, and (3) the nets are rerouted in the next iteration. This process continues until all nets are routed or some stopping condition is reached.

The edge cost $cost(e)$ is increased according to the *edge congestion* $\varphi(e)$, defined as the total number of nets passing through e divided by the capacity of e.

$$\varphi(e) = \frac{\eta(e)}{\sigma(e)}$$

If e is uncongested, i.e., $\varphi(e) \leq 1$, then $cost(e)$ does not change. If e is congested, i.e., $\varphi(e) > 1$, then $cost(e)$ is increased so as to penalize nets that use e in subsequent iterations. In NCR, $cost(e)$ only increases or remains the same.[7] In practice, the edge costs are updated after initial routing and after every subsequent routing iteration. As such, the number of different routes having identical costs is reduced for each net, making net ordering less important for global routing.

The rate $\Delta cost(e)$ at which $cost(e)$ grows must be controlled. If $\Delta cost(e)$ is too high, then entire groups of nets will be simultaneously pushed away from one edge and toward another. This can cause the nets' routes to bounce back and forth between edges, leading to longer runtime and longer routes, and possibly jeopardizing successful routing. On the other hand, if $\Delta cost(e)$ is too low, then many more iterations will be required to route all nets without violation, causing an increase in runtime. Ideally, the rate of cost increase should be gradual so that a fraction of nets will be routed differently in each iteration. In different routers, this rate has been modeled by linear functions [5.3], dynamically changing logistic functions [5.16], and exponential functions with a slowly-growing constant [5.1][5.21]. In practice, well-tuned NCR-based routers can effectively reduce congestion while maintaining low wirelength, defined as the routed length plus the number of vias (Fig. 5.23).

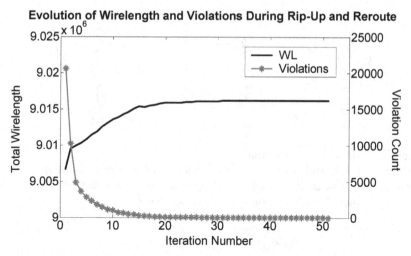

Fig. 5.23 Progression of wirelength and violation count during the rip-up and reroute stage.

---

[7] If $cost(e)$ were to decrease, then nets previously penalized for using e no longer have that cost, and the effort to route these nets in previous iterations will be wasted since those nets will then use the same edges as before.

Chapter 5 Exercises

Exercise 1: Steiner Tree Routing
Given the six-pin net on the routing grid (right).
(a) Mark all Hanan points and draw the MBB.
(b) Generate the RSMT using the heuristic in
      Sec. 5.6.1. Show all intermediate steps.
(c) Determine the degree of each Steiner point
      in the RSMT for the given net.
(d) Determine the maximum number of Steiner
      points that a three-pin RSMT can have.

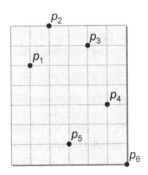

Exercise 2: Global Routing in a Connectivity Graph
Two nets A and B and the connectivity graph with capacities are given. Determine
whether this placement is routable. If the placement is not routable, explain why. In
either case, routable or unroutable, calculate the remaining capacities after both nets
have been routed.

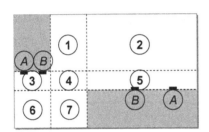
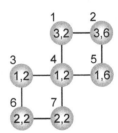

Exercise 3: Dijkstra's Algorithm
For the graph with weights (w_1, w_2) shown below, use Dijkstra's algorithm to find a
minimum-cost path from the starting node $s = a$ to the target node $t = i$. Generate the
tables for Groups 2 and 3 as in the example of Sec. 5.6.3.

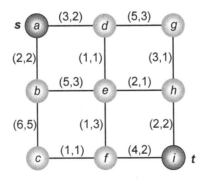

Exercise 4: ILP-Based Global Routing
Modify the example given in Sec. 5.7.1 by disallowing Z-shape routes. Give the full ILP instance and state whether it is feasible, i.e., has a valid solution. If a solution exists, then illustrate the routes on the grid. Otherwise, explain why no solution exists.

Exercise 5: Shortest Path with A* Search
Modify the example illustrated in Fig. 5.19 by *removing* one obstacle. Number the nodes searched as in Fig. 5.19(b).

Exercise 6: Rip-Up and Reroute
Consider rip-up and reroute on an $m \times m$ grid with n nets. Estimate the required memory usage. Choose from the following.

$$O(m^2) \qquad O(m^2 + n) \qquad O(m^2 \cdot n^2)$$
$$O(m^2 \cdot n) \qquad O(n^2) \qquad O(m \cdot n) \qquad O(m \cdot n^2)$$

Chapter 5 References

[5.1]   S. Batterywala, N. Shenoy, W. Nicholls and H. Zhou, "Track Assignment: A Desirable Intermediate Step Between Global and Detailed Routing", *Proc. Intl. Conf. on CAD*, 2002, pp. 59-66.

[5.2]   Y.-J. Chang, Y.-T. Lee and T.-C. Wang, "NTHU-Route 2.0: A Fast and Stable Global Router", *Proc. Intl. Conf. on CAD*, 2008, pp. 338-343.

[5.3]   H.-Y. Chen, C.-H. Hsu and Y.-W. Chang, "High-Performance Global Routing with Fast Overflow Reduction", *Proc. Asia and South Pacific Design Autom. Conf.*, 2009, pp. 582-587.

[5.4]   M. Cho and D. Pan, "BoxRouter: A New Global Router Based on Box Expansion", *IEEE Trans. on CAD* 26(12) (2007), pp. 2130-2143.

[5.5]   C. Chu and Y. Wong, "FLUTE: Fast Lookup Table Based Rectilinear Steiner Minimal Tree Algorithm for VLSI Design", *IEEE Trans. on CAD* 27(1) (2008), pp. 70-83.

[5.6]   E. Dijkstra, "A Note on Two Problems in Connexion With Graphs", *Num. Math.* 1 (1959), pp. 269-271.

[5.7]   GLPK, gnu.org/software/glpk.

[5.8]   M. Hanan, "On Steiner's Problem with Rectilinear Distance", *SIAM J. on App. Math.* 14(2) (1966), pp. 255-265.

[5.9]   P. E. Hart, N. J. Nilsson and B. Raphael, "A Formal Basis for the Heuristic Determination of Minimum Cost Paths", *IEEE Trans. on Sys. Sci. and Cybernetics* 4(2) (1968), pp. 100-107.

[5.10] J.-M. Ho, G. Vijayan and C. K. Wong, "New Algorithms for the Rectilinear Steiner Tree Problem", *IEEE Trans. on CAD* 9(2) (1990), pp. 185-193.

[5.11] J. Hu, J. Roy and I. Markov, "Completing High-Quality Global Routes", *Proc. Intl. Symp. on Phys. Design*, 2010, pp. 35-41.

[5.12] J. Hu, J. Roy and I. Markov, "Sidewinder: A Scalable ILP-Based Router", *Proc. Sys. Level Interconnect Prediction*, 2008, pp. 73-80.

[5.13] ILOG CPLEX, www.cplex.com.

[5.14] E. S. Kuh and T. Ohtsuki, "Recent Advances in VLSI Layout", *Proc. IEEE* 78(2) (1990), pp. 237-263.

[5.15] L. McMurchie and C. Ebeling, "Pathfinder: A Negotiation-Based Performance-Driven Router for FPGAs", *Proc. Intl. Symp. on FPGAs*, 1995, pp. 111-117.

[5.16] M. Moffitt, "MaizeRouter: Engineering an Effective Global Router", *IEEE Trans. on CAD* 27(11) (2008), pp. 2017-2026.

[5.17] MOSEK, www.mosek.com.

[5.18] G.-J. Nam, C. Sze and M. Yildiz, "The ISPD Global Routing Benchmark Suite", *Proc. Intl. Symp. on Phys. Design*, 2008, pp. 156-159.

[5.19] R. C. Prim, "Shortest Connection Networks and Some Generalizations", *Bell Sys. Tech. J.* 36(6) (1957), pp. 1389-1401.

[5.20] H.-J. Rothermel and D. Mlynski, "Automatic Variable-Width Routing for VLSI", *IEEE Trans. on CAD* 2(4) (1983), pp. 271-284.

[5.21] J. A. Roy and I. L. Markov, "High-Performance Routing at the Nanometer Scale", *IEEE Trans. on CAD* 27(6) (2008), pp. 1066-1077.

[5.22] Y. Xu, Y. Zhang and C. Chu, "FastRoute 4.0: Global Router with Efficient Via Minimization", *Proc. Asia and South Pac. Design Autom. Conf.*, 2009, pp. 576-581.

[5.23] Y. Zheng, Y. Xu and C. Chu, "FastRoute 3.0: A Fast and High Quality Global Router Based on Virtual Capacity", *Proc. Intl. Conf. on CAD*, 2008, pp. 344-349.

Chapter 6

Detailed Routing

6

6

6 Detailed Routing

Recall from Chap. 5 that the layout region is represented by a coarse grid consisting of *global routing cells* (gcells) or more general *routing regions* (channels, switchboxes) during global routing. Afterward, each net undergoes *detailed routing*.

The objective of detailed routing is to assign route segments of signal nets to specific routing tracks, vias, and metal layers in a manner consistent with given global routes of those nets. These route assignments must respect all design rules.

Each gcell is orders of magnitude smaller than the entire chip, e.g., 10 × 10 routing tracks, regardless of the actual chip size. As long as the routes remain properly connected across all neighboring gcells, the detailed routing of one gcell can be performed independently of the routing of other gcells. This facilitates an efficient divide-and-conquer framework and also enables parallel algorithms. Thus, detailed routing runtime can (theoretically) scale linearly with the size of the layout. Traditional detailed routing techniques are applied within routing regions, such as channels (Sec. 6.3) and switchboxes (Sec. 6.4). For modern designs, *over-the-cell* (*OTC*) routing (Sec. 6.5) allows wires to be routed over standard cells. Due to technology scaling, modern detailed routers must account for manufacturing rules and the impact of manufacturing faults (Sec. 6.6).

6.1 Terminology

Channel routing is a special case of detailed routing where the connections between terminal pins are routed within a routing region (channel) that has no obstacles. The pins are located on *opposite sides* of the channel (Fig. 6.1, left). By convention, the channel is oriented horizontally – pins are on the top and bottom of the channel. In row-based layouts, in a given block, the routing channels typically have uniform *channel width*. In gate-array and standard-cell circuits that use more than three layers of metal, *channel height*, the number of routing tracks between the top and bottom boundaries of the channel, is also uniform.

Switchbox routing is performed when pin locations are given on all four sides of a fixed-size routing region (switchbox, Fig. 6.1, right). This makes the detailed routing significantly more difficult than in channel routing. Switchbox routing is further discussed in Sec. 6.4.

OTC (over-the-cell) routing uses additional metal tracks, e.g., on Metal3 and Metal4, that are not obstructed by cells, allowing routes to cross cells and channels. An example is shown in Fig. 6.2. OTC routing can use only the metal layers and tracks that the cells do not occupy. When the cells utilize only the polysilicon and Metal1 layers, routing can be performed on the remaining metal layers (Metal2, Metal3, etc.) as well as unused Metal1 resources. OTC routing is further discussed in Sec. 6.5.

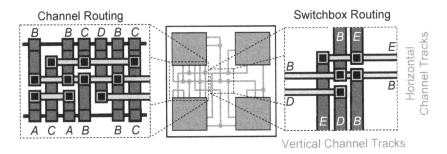

Fig. 6.1 Example of two-layer channel and switchbox routing. The pins of each net *A-D* are all positioned perpendicular to the channel. Each layer has a preferred direction – one layer has only horizontal tracks and one layer has only vertical tracks. Vias are used if routing a net requires both horizontal and vertical tracks.

In classical channel routing, the routing area is a rectangular grid (Fig. 6.3) with pin locations on top and bottom boundaries. The pins are located on the vertical grid lines or *columns*. The channel height depends on the number of *tracks* that are needed to route all the nets. In two-layer routing, one layer is reserved exclusively for horizontal tracks while the other is reserved for vertical tracks. The preferred direction of each routing layer is determined by the floorplan and the orientation of the standard-cell rows. In a horizontal cell row, polysilicon (transistor gate) segments are typically vertical (*V*) and Metal1 segments are horizontal (*H*). The metal layers' preferred directions then alternate between *H* and *V*. To connect to a cell pin that is on the Metal1 layer, the router will drop one or more vias from a Metal2 routing segment.

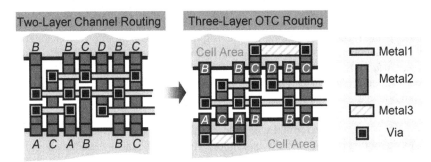

Fig. 6.2 An example of OTC routing. Note that fewer routing tracks are required if some nets are routed over the cell area.

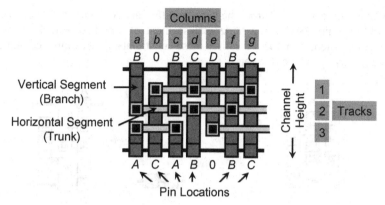

Fig. 6.3 Terminology related to channel routing for a horizontal channel. The columns are the vertical grid lines while the tracks are the horizontal grid lines.

Given a channel, its upper and lower boundaries are each defined by a vector of net IDs, denoted as *TOP* and *BOT*, respectively. Here, each column is represented by two net IDs (Fig. 6.3), one from the top channel boundary and one from the bottom channel boundary. Unconnected pins are given a net ID of 0. In the example of Fig. 6.3, *TOP* = [*B* 0 *B C D B C*] and *BOT* = [*A C A B* 0 *B C*].

A *horizontal constraint* exists between two nets if their horizontal segments overlap when placed on the same track. The example in Fig. 6.4 includes one horizontal and one vertical routing layer, with nets *B* and *C* being horizontally constrained. If the two nets' horizontal segments do not overlap, then they can both be assigned to the same track and are horizontally unconstrained, e.g., nets *A* and *B* in Fig. 6.4.

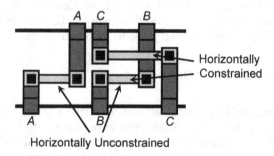

Fig. 6.4 Example of horizontally constrained and unconstrained nets. Nets *B* and *C* are horizontally constrained and thus require different horizontal tracks.

A *vertical constraint* exists between two nets if they have pins in the same column. In other words, the vertical segment coming from the top must "stop" within a short distance so that it does not overlap with the vertical segment coming from the bottom in the same column (Fig. 6.5). If each net is assigned to a single horizontal track, then the horizontal segment of a net from the top must be placed above the horizontal segment of a net from the bottom in the same column. In Fig. 6.5(a), this vertical constraint assigns the horizontal segment of net A to a track above the horizontal segment of net B. To satisfy these constraints, at least three columns are required to "uncross" the two nets.

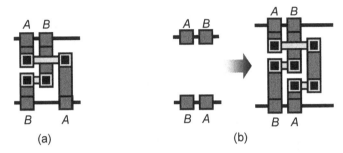

Fig. 6.5 Examples of vertically constrained nets. (a) Nets A and B do not have a vertical conflict. (b) Nets A and B have a vertical conflict. The two nets can be routed by splitting the vertical segment of one net and using an additional third track.

Although a vertical constraint implies a horizontal constraint, the converse is not necessarily true. However, both types of constraints must be satisfied when assigning segments in a channel.

6.2 Horizontal and Vertical Constraint Graphs

The relative positions of nets in a channel routing instance, encoded with horizontal and vertical constraints, can be modeled by *horizontal* and *vertical constraint graphs*, respectively. These graphs are used to (1) initially predict the minimum number of tracks that are required and (2) detect potential routing conflicts.

▶ 6.2.1 Horizontal Constraint Graphs

Zone representation. In a channel, all horizontal wire segments must span at least the leftmost and rightmost pins of their respective nets. Let $S(col)$ denote the set of nets that pass through column col. In other words, $S(col)$ contains all nets that either (1) are connected to a pin in column col or (2) have pin connections to both the left and right of col. Since horizontal segments cannot overlap, each net in $S(col)$ must be assigned to a different track in column col. Only a subset of all columns is needed to describe the entire channel. If there exist columns i and j such that $S(i)$ is a subset of $S(j)$, then $S(i)$ can be ignored since it imposes fewer constraints on the routing solution than $S(j)$. In Fig. 6.6, every $S(col)$ is a subset of at least one of $S(c)$, $S(f)$, $S(g)$ or $S(i)$. Furthermore, the *maximal* columns c, f, g and i comprise a minimal set of columns with this property. Note that these columns together contain all the nets.

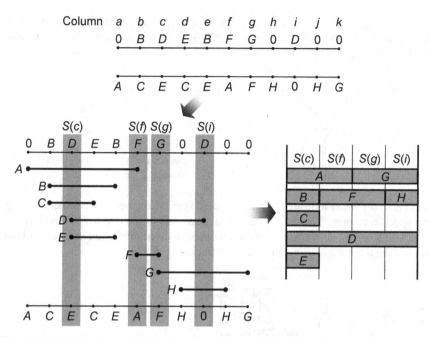

Fig. 6.6 A channel routing problem (top) and its corresponding zone representation (right). Only the maximal columns are shown.

Graphical representation. The nets within the channel can also be represented by a *horizontal constraint graph HCG(V,E)*, where nodes $v \in V$ correspond to the nets of the netlist and an undirected edge $e(i,j) \in E$ exists between nodes i and j if the corresponding nets are both elements of some set $S(col)$. In other words, $e \in E$ if the corresponding nets are horizontally constrained. Fig. 6.7 illustrates the HCG for the channel routing instance of Fig. 6.6. A lower bound on the number of tracks required by the channel routing can be found from either the HCG or the zone representation. This lower bound is given by the maximum cardinality of any $S(col)$.

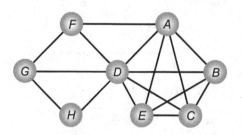

Fig. 6.7 The HCG for the channel routing instance of Fig. 6.6. With either the HCG or the zone representation, a lower bound on the number of tracks is five.

► 6.2.2 Vertical Constraint Graphs

Vertical constraints are represented by a *vertical constraint graph VCG(V,E)*. A node $v \in V$ represents a net. A directed edge $e(i,j) \in E$ connects nodes i and j if net i

must be located above net *j*. However, an edge that can be derived by transitivity is not included. For instance, in Fig. 6.8, edge (*B*,*C*) is not included because it can be derived from edges (*B*,*E*) and (*E*,*C*).

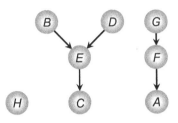

Fig. 6.8 The VCG for the channel routing instance of Fig. 6.6.

A cycle in the VCG indicates a conflict where vertical segments of two nets overlap at a specific column. That is, the horizontal segments of the two nets would have to be simultaneously above and below each other. This contradiction can be resolved by splitting the net and using an additional track (Fig. 6.9).

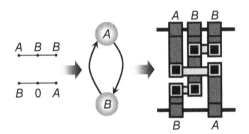

Fig. 6.9 A channel routing problem (left), its VCG with a cycle or conflict (center), and a possible solution using net splitting and an additional third track (right).

If a cycle occurs in a maximum-cardinality set *S(col)* of the HCG, the lower bound on the number of required tracks, based on the number of nets in *S(col)*, is no longer tight. Nets must now be split and the minimum number of required tracks must be adjusted to account for both HCG and VCG conflicts.

Example: Vertical and Horizontal Constraint Graphs
Given: channel routing instance (right).
Task: find the horizontal constraint graph (HCG) and the vertical constraint graph (VCG) of nets *A-F*.

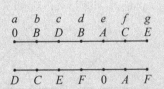

Column	a	b	c	d	e	f	g
	0	B	D	B	A	C	E

| D | C | E | F | 0 | A | F |

Solution:
TOP and *BOT* vectors: *TOP* = [0 *B D B A C E*], *BOT* = [*D C E F* 0 *A F*]
Determine *S(col)* for *col* = *a* ... *g*.

$$S(a) = \{D\} \qquad S(b) = \{B,C,D\} \qquad S(c) = \{B,C,D,E\} \quad S(d) = \{B,C,E,F\}$$
$$S(e) = \{A,C,E,F\} \quad S(f) = \{A,C,E,F\} \quad S(g) = \{E,F\}$$

Find the maximal *S(col)*.
S(a) = {*D*} and *S(b)* = {*B,C,D*} are both subsets of *S(c)* = {*B,C,D,E*}.
S(f) = {*A,C,E,F*} and *S(g)* = {*E,F*} are both subsets of *S(e)* = {*A,C,E,F*}.
Maximal *S(col)*: *S(c)* = {*B,C,D,E*}   *S(d)* = {*B,C,E,F*}   *S(e)* = {*A,C,E,F*}

Find the HCG and VCG.

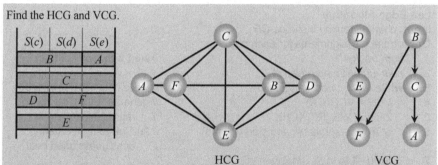

$S(c)$	$S(d)$	$S(e)$
	B	A
	C	
D	F	
	E	

HCG   VCG

Since there are no cycles in the VCG, no *net splitting* is required (Sec. 6.2.2). Thus, each net only needs one horizontal segment for routing.

The track assignment is based on both the VCG and the HCG. For instance, based on the VCG, net D is assigned to the topmost track. The other net that is on the top level in the VCG (net B) is assigned to another track due to the HCG.

The HCG determines a lower bound on the number of required tracks. Since there are no cycles (conflicts) in the VCG, the minimum number of tracks is equal to the cardinality of the largest $S(col)$, here being $|S(c)| = |S(d)| = |S(e)| = 4$.

6.3 Channel Routing Algorithms

Channel routing seeks to minimize the number of tracks required to complete routing. In gate-array designs, channel height is typically fixed, and algorithms are designed to pursue 100% routing completion.

► 6.3.1 Left-Edge Algorithm

An early channel routing algorithm was developed by *Hashimoto* and *Stevens* [6.8]. Their simple and widely used *left-edge* heuristic, based on the VCG and the zone representation, greedily maximizes the usage of each track. The former identifies the assignment order of nets to tracks, and the latter determines which nets may share the same track. Each net uses only one horizontal segment (trunk).

The left-edge algorithm works as follows. Start with the topmost track (line 1). For all unassigned nets *nets_unassigned* (line 3), generate the VCG and the zone representation (lines 4-5). Then, in left-to-right order (line 6), for each unassigned net *n*, assign it to the current track if (1) *n* has no predecessors in the VCG and (2) it does not cause a conflict with any nets that have been previously assigned (lines 7-11). Once *n* has been assigned, remove it from *nets_unassigned* (line 12). After all unassigned nets have been considered, increment the track index (line 13). Continue this process until all nets have been assigned to routing tracks (lines 3-13).

Left-Edge Algorithm
Input: channel routing instance *CR*
Output: track assignments for each net

1.	*curr_track* = 1	// start with topmost track
2.	*nets_unassigned* = *Netlist*	
3.	**while** (*nets_unassigned* != Ø)	// while nets still unassigned
4.	VCG = VCG(*CR*)	// generate VCG and zone
5.	ZR = ZONE_REP(*CR*)	// representation
6.	SORT(*nets_unassigned*,start column)	// find left-to-right ordering
		// of all unassigned nets
7.	**for** (*i* =1 to |*nets_unassigned*|)	
8.	*curr_net* = *nets_unassigned*[*i*]	
9.	**if** (PARENTS(*curr_net*) == Ø &&	// if *curr_net* has no parent
10.	(TRY_ASSIGN(*curr_net*,*curr_track*))	// and does not cause
		// conflicts on *curr_track*,
11.	ASSIGN(*curr_net*,*curr_track*)	// assign *curr_net*
12.	REMOVE(*nets_unassigned*,*curr_net*)	
13.	*curr_track* = *curr_track* + 1	// consider next track

The left-edge algorithm finds a solution with the minimum number of tracks, or the maximum cardinality of *S*(*col*), if there are no cycles in the VCG. *Yoshimura* [6.18] enhanced track selection by incorporating net length in the VCG. *Yoshimura* and *Kuh* [6.19] improved track utilization by net splitting before constructing the VCG.

Example: Left-Edge Algorithm
Given: channel routing instance (right).

Task: use the left-edge algorithm to route nets *A-J* in the channel.

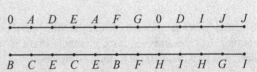

Solution:
curr_track = 1, VCG and zone representation:

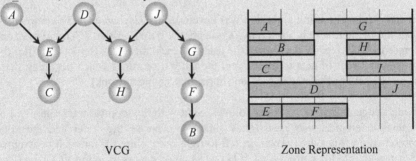

VCG   Zone Representation

Nets *A*, *D* and *J* have no predecessors in the VCG and can be assigned to *curr_track* = 1. Net *A* is assigned first because it is the leftmost. Net *D* can no longer be assigned to *curr_track* because of a conflict with net *A*. However, net *J* can be assigned to *curr_track*, since there is no conflict. Remove nets *A* and *J* from the VCG and the zone representation.
curr_track = 2, VCG and zone representation:

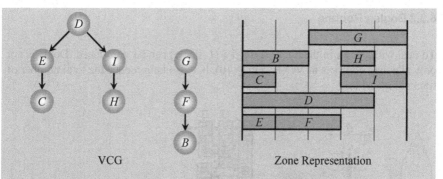

VCG   Zone Representation

Nets *D* and *G* have no predecessors in the VCG and can be assigned to *curr_track* = 2. Net *D* is assigned first because it is the leftmost. Net *G* can no longer be assigned to *curr_track* because of a conflict with net *D*. Remove net *D* from the VCG and the zone representation.

curr_track = 3, VCG and zone representation:

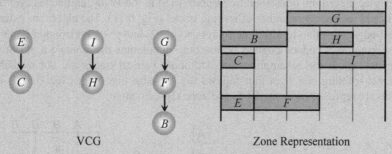

VCG   Zone Representation

Nets *E*, *G* and *I* have no predecessors in the VCG and can be assigned to *curr_track* = 3. Net *E* is assigned first because it is the leftmost. Net *G* is assigned to *curr_track* because it does not conflict with any nets. Net *I* cannot be assigned because of a conflict with net *G*. Remove nets *E* and *G* from the VCG and the zone representation.

curr_track = 4 and *curr_track* = 5 are similar to previous iterations. Nets *C*, *F* and *I* are assigned to track 4. Nets *B* and *H* are assigned to track 5.

Routed channel:

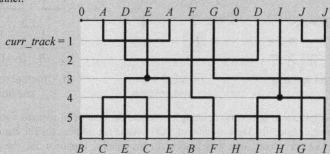

▶ 6.3.2 Dogleg Routing

To deal with cycles in the VCG, a *dogleg* (*L*-shape) can be introduced. Doglegs not only alleviate conflicts in VCGs (Fig. 6.10), but also help reduce the total number of tracks (Fig. 6.11).

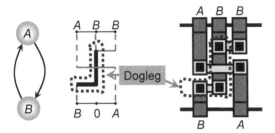

Fig. 6.10 A dogleg is introduced to route around the conflict for net *B*.

The dogleg algorithm, developed by *Deutsch* [6.6] in the 1970s, eliminates cycles in VCGs and reduces the number of routing tracks (Fig. 6.11). The algorithm extends the left-edge algorithm by splitting p-pin nets ($p > 2$) into $p - 1$ horizontal segments. Net splitting to introduce doglegs occurs only in columns that contain a pin of the given net, under the assumption that additional vertical tracks are not available. After net splitting, the algorithm follows the left-edge algorithm (Sec. 6.3.1). The subnets are represented by the VCG and zone representation.

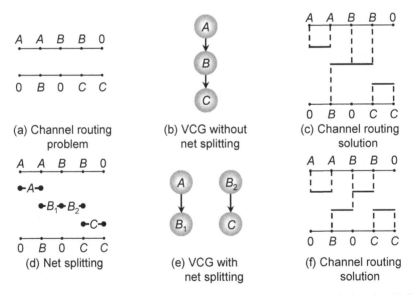

Fig. 6.11 Example showing how net splitting can reduce the number of required tracks. (a) The channel routing instance. (b) The conventional VCG without net splitting. (c) The channel routing solution without net splitting uses three tracks. (d) Net splitting applied to the channel routing solution from (c). (e) The new VCG after net splitting. (f) The new channel routing solution with net splitting uses only two tracks.

Example: Dogleg Left-Edge Algorithm

Given: channel routing instance (right).

Task: use the dogleg left-edge algorithm to route the nets A-D in the channel.

Column	a	b	c	d	e	f
	C	D	0	D	A	A

| | B | B | C | 0 | C | D |

Solution:

Split the nets, find $S(col)$ and determine the zone representation. Note: subnets of a net can be placed on the same track regardless of overlap in the zone representation.

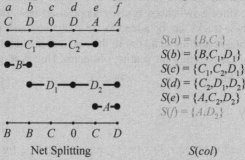

$$S(a) = \{B, C_1\}$$
$$S(b) = \{B, C_1, D_1\}$$
$$S(c) = \{C_1, C_2, D_1\}$$
$$S(d) = \{C_2, D_1, D_2\}$$
$$S(e) = \{A, C_2, D_2\}$$
$$S(f) = \{A, D_2\}$$

Net Splitting $S(col)$

Find the VCG.

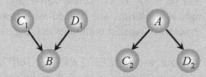

Track assignment:

curr_track = 1

Consider nets C_1, D_1 and A. Assign net C_1 first, since it is the leftmost net in the zone representation. Of the remaining nets, only net A does not cause a conflict. Therefore, assign net A to *curr_track*. Remove nets C_1 and A from the VCG.

curr_track = 2

Consider nets D_1, C_2 and D_2. Assign net D_1 first, since it is the leftmost net in the zone representation. Of the remaining nets, only net D_2 does not cause a conflict. Therefore, assign net D_2 to *curr_track*. Remove nets D_1 and D_2 from the VCG.

curr_track = 3

Consider nets B and C_2. Assign net B first, since it is the leftmost net in the zone representation. Net C_2 does not cause a conflict. Therefore, assign net C_2 to *curr_track*.

Routed channel:

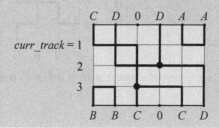

6.4 Switchbox Routing

Recall that switchboxes have fixed dimensions and include pin connections on all four sides. Switchbox routing seeks to connect all pins in each set with identical labels. Nets can be routed on specific horizontal tracks or vertical columns, and are allowed to cross when tracks and columns are on multiple layers. Compared to channel routing, pins on all four sides lead to a larger number of crossings and greater complexity, since a switchbox router cannot insert new tracks. If switchbox routing fails, then new tracks can be added to the switchbox and adjacent channels, and switchbox routing is attempted again. In this section, the algorithm descriptions are simplified by not accounting for routing obstacles. However, all algorithms can be extended accordingly.

▶ 6.4.1 Terminology

A *switchbox* is defined as an $(m + 1) \times (n + 1)$ region with $0 \ldots (m + 1)$ columns and $0 \ldots (n + 1)$ rows. The 0th and $(m + 1)^{th}$ *columns* are the left and right borders of the switchbox, and the 0th and $(n + 1)^{th}$ *rows* are the lower and upper borders of the switchbox. The 1st through m^{th} columns are labeled with lowercase letters, e.g., a and b; the 1st through n^{th} tracks are labeled with numbers, e.g., 1 and 2.

A switchbox is defined by four vectors *LEFT*, *RIGHT*, *TOP* and *BOT*, where they each respectively define the pin ordering on the left, right, top, and bottom borders. Since pins are located on all four borders, the usability of these borders for routing is severely restricted.

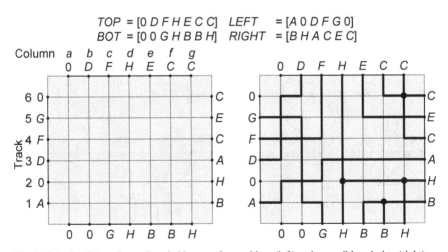

Fig. 6.12 An 8×7 ($m = 7, n = 6$) switchbox routing problem (left) and a possible solution (right).

► 6.4.2 Switchbox Routing Algorithms

Algorithms for switchbox routing can be derived from channel routing algorithms. *Luk* [6.14] extended a greedy channel router by *Rivest* and *Fiduccia* [6.16] to propose a switchbox routing algorithm with the following key improvements.

1. Pin assignments are made on all four sides.
2. A horizontal track is automatically assigned to a pin on the left.
3. *Jogs* are used for the top and bottom pins as well as for horizontal tracks connecting to the rightmost pins.

The performance of this algorithm is similar to that of the greedy channel router from [6.16], but it does not guarantee full routability because the switchbox dimensions are fixed.

Example: Switchbox Routing
Given: the 8 × 7 ($m = 7$, $n = 6$) switchbox routing instance of Fig. 6.12.

$TOP = [0\ D\ F\ H\ E\ C\ C]$ $LEFT\ \ = [A\ 0\ D\ F\ G\ 0]$
$BOT = [0\ 0\ G\ H\ B\ B\ H]$ $RIGHT\ = [B\ H\ A\ C\ E\ C]$

Task: route nets *A-H* within the switchbox.

Solution:
Column *a*: Assign net *A* to track 2. Assign net *D* to track 6. Extend nets *A* (track 2), *F* (track 4), *G* (track 5) and *D* (track 6).
Column *b*: Connect the top pin *D* to net *D* on track 6. Assign net *G* with track 1. Extend nets *G* (track 1), *A* (track 2) and *F* (track 4).
Column *c*: Connect the top pin *F* to net *F* on track 4. Connect the bottom pin *G* to net *G* on track 1. Assign net *A* to track 3. Extend net *A* (track 3).
Column *d*: Connect the bottom pin *H* to the top pin *H*. Extend nets *H* (track 2) and *A* (track 3).
Column *e*: Connect the bottom pin *B* with track 1. Connect the top pin *E* with track 5. Extend nets *B* (track 1), *H* (track 2), *A* (track 3) and *E* (track 5).
Column *f*: Connect the bottom pin *B* to net *B* on track 1. Connect the top pin *C* with track 6. Extend nets *B* (track 1), *H* (track 2), *A* (track 3), *E* (track 5) and *C* (track 6).
Column *g*: Connect the bottom pin *H* with net *H* on track 2. Connect the top pin *C* with net *C* on track 6. Assign net *C* to track 4. Extend the nets on tracks 1, 2, 3, 4, 5 and 6 to their corresponding pins.

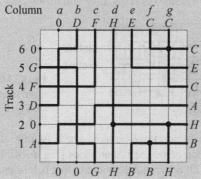

Ousterhout et al. developed a channel and switchbox router that accounts for obstacles such as pre-routed nets [6.15] based on the greedy channel router in [6.16]. *Cohoon* and *Heck* developed the switchbox router *BEAVER*, which accounts for vias and minimizes total routing area [6.3]. It allows additional flexibility for preferred routing directions on individual layers. *BEAVER* uses the following strategies – (1) corner-routing, where a horizontal and vertical segment form a *bend*, (2) line-sweep routing for simple connections and straight segments, (3) thread-routing where any type of connection can be made, and (4) layer assignment. *BEAVER* outperforms previous academic routers in terms of routing area and via count.

Another well-known switchbox router, *PACKER*, developed by *Gerez* and *Herrmann* in 1989 [6.7] has three major steps. First, each net is routed independently, ignoring capacity constraints. Second, any remaining conflicts are resolved using *connectivity-preserving local transformations* (*CPLT*). Third, the net segments are then locally modified (rerouted) to alleviate routing congestion.

6.5 Over-the-Cell Routing Algorithms

Most routing algorithms in Secs. 6.3-6.4 have dealt primarily with two-layer routing. However, modern standard-cell designs have more than two layers, so these algorithms must be extended accordingly. One commonly used strategy proceeds as follows. The cells between the channels are placed *back-to-back* or without routing channels. Cells predominantly use only Poly and Metal1 for internal routing. Higher metal layers, e.g., Metal2 and Metal3, are not obstructed by standard cells and are typically used for *over-the-cell* (*OTC*) routing. These metal layers are usually represented by a coarse routing grid made up of gcells. The nets are globally routed as Steiner trees (Sec. 5.6.1) and then detail-routed (Secs. 6.3-6.4).

In an alternate approach, channels are created between the cells, but are limited to the internal cell layers such as Poly and Metal1. Routing is generally performed on higher metal layers, such as Metal2 and Metal3. Since standard cells do not form routing obstacles at these higher layers, the concept of a routing channel is irrelevant in this context. Therefore, routing is performed on the entire chip area rather than in individual channels or switchboxes (Figs. 6.13 and 6.14).

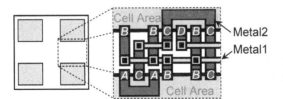

Fig. 6.13 OTC routing with two metal layers. For another example, see Fig. 1.7.

For designs with more than three layers, gcells within the layout region are decomposed and stretched across cell boundaries (Fig. 6.14). In addition, the power (*VDD*) and ground (*GND*) nets require an alternating cell orientation (Fig. 1.7).

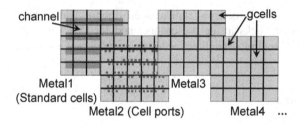

channel

gcells

Metal1
(Standard cells)   Metal3

Metal2 (Cell ports)   Metal4 ...

Fig. 6.14 Routing (metal) layers of a design partitioned into global routing cells (gcells). Standard-cell regions in Metal1 are colored using dark gray.

OTC routing sometimes coexists with channel routing. For example, IP blocks typically block routing on several lowest metal layers, so the space between IP blocks can be broken down as channels or switchboxes. Furthermore, FPGA fabrics typically use very few metal layers to reduce manufacturing cost, and therefore cluster programmable interconnect into channels between *logic elements*. FPGAs can also include pre-designed multipliers, *digital signal processing* (*DSP*) blocks and memories, which use OTC routing. Recent FPGAs also include *express-wires* laid out on higher metal layers that may cross logic elements.

▶ **6.5.1 OTC Routing Methodology**

OTC routing is performed in three steps – (1) select nets that will be routed outside the channel, (2) route these nets in the OTC area, and (3) route the remaining nets within the channel. *Cong* and *Liu* [6.4] solved steps 1 and 2 optimally in $O(n^2)$ time, where n is the number of nets.

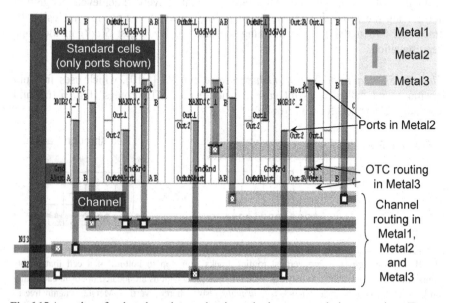

Fig. 6.15 A portion of a three-layer layout showing only the routes and pin connections (Tanner Research, Inc.). Metal2 segments are predominantly vertical. Metal1 and Metal3 segments are primarily horizontal; Metal3 is wider than other segments (excluding the ground stripe on the left).

Fig. 6.15 shows a fragment from a three-layer standard-cell design. The Metal3 layer is used both for horizontal tracks inside the channel and for OTC routing. While early applications of OTC routing used three metal layers, modern ICs use six or more metal layers.

▶ 6.5.2 OTC Routing Algorithms

This section discusses several advanced OTC routing algorithms. Further details can be found in respective publications. The Chameleon OTC router [6.2], developed by *Braun et al.*, first generates topologies for nets in either two or three metal layers to minimize the total routing area, and then assigns the net segments to tracks. The key innovation of Chameleon is its ability to take into account technology parameters on a per-layer basis. For example, designers can specify the wire widths and the minimum distance between each wire segment for each layer.

Cong, *Wong* and *Liu* in [6.5] chose a different approach to OTC routing. All the nets are first routed in two layers, and then mapped onto three layers. Shortest-path and planar-routing algorithms are used to map two-layer routes to three-layer routes with minimum wiring area. The authors assume an *H-V-H* three-layer model whereby the first and third layers host horizontal tracks, and the second layer hosts vertical tracks. The approach can be extended to four-layer routing.

Ho et al. [6.9] developed another OTC routing technique that routes nets greedily, using a set of simple heuristics which are applied iteratively. It achieved recognition for solving the *Deutsch Difficult Example* [6.6] channel routing instance using 19 tracks on two metal layers, matching the size of the largest $S(col)$ in the zone representation.

Holmes, *Sherwani* and *Sarrafzadeh* presented *WISER* [6.10], which uses free pin positions and cell area to increase the number of OTC routes (Fig. 6.16), and carefully selects nets for OTC routing.

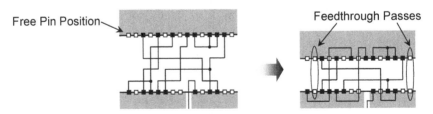

Fig. 6.16 WISER uses free pin positions to reduce the number of channel tracks from four to two [6.10]. *Feedthroughs* are free vertical routing tracks from top to bottom of a cell that enable connections between adjacent channels. When unused pin locations are located opposite each other on the top and bottom of a channel, *feedthroughs* result whereby vertical Metal2 is again free to make a connection between adjacent channels. An unused pin location means that there is no Metal1 feature (pin) at that location inside the standard cell, and hence no need for Metal2 to connect to any standard-cell pin. Hence, the Metal2 resource is available.

6.6 Modern Challenges in Detailed Routing

The need for low-cost, high-performance and low-power ICs has driven technology scaling since the 1960s [6.11]. An important aspect of modern technology scaling is the use of wires of different widths on different metal layers. In general, wider wires on higher metal layers allow signals to travel much faster than on thinner wires on lower metal layers. This helps to recover some benefits from scaling in terms of performance, but at the cost of fewer routing tracks. Thicker wires are typically used for clock (Sec. 7.5) and supply routing (Sec. 3.7), as well as for global interconnect.

Manufacturers today use different configurations of metal layers and widths to accommodate high-performance designs. However, such a variety of routing resources makes detailed routing more challenging. Vias connecting wires of different widths inevitably block additional routing resources on the layer with the smaller wire pitch. For example, layer stacks in some IBM designs for 130 nm-32 nm technologies are illustrated in Fig. 6.17 [6.1]. Wires on layers M have the smallest possible width λ, while the wires on layers C, B, E, U and W are wider – 1.3λ, 2λ, 4λ, 10λ, and 16λ, respectively. The 90 nm technology node was the first to introduce different metal layer thicknesses, with thinner wires on the top two layers. Today's 32 nm metal layer stacks often incorporate four to six distinct wire thicknesses. Advanced lithography techniques used in manufacturing lead to stricter enforcement of *preferred routing direction* on each layer.

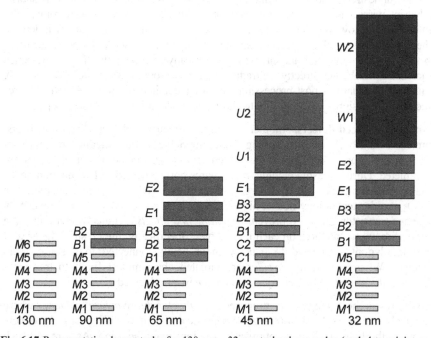

Fig. 6.17 Representative layer stacks for 130 nm - 32 nm technology nodes (scaled to minimum feature size at each technology node).

Semiconductor manufacturing yield is a key concern in detailed routing. To safeguard against manufacturing defects, *via doubling* and *non-tree routing* insert redundant vias and wiring segments as backups in case an electrical connection is lost. At advanced technology nodes, manufacturability constraints (design rules) become more restrictive and notably complicate detailed routing. For example, design rules specify minimum allowed spacing between wires and vias depending on their widths and proximity to wire corners. More recent spacing rules take into account multiple neighboring polygons. *Forbidden pitch* rules prohibit routing wires at certain distances apart, but allows smaller or greater spacings.

Via defects. Recall that a (single) via connects two wires on different metal layers. However, vias can be misaligned during manufacturing, and are susceptible to *electromigration effects* during the chip's lifetime [6.13]. A partially failing via with increased resistance may cause timing violations in the circuit. A via that has failed completely may disconnect a net, altering the circuit's function. To protect against via failures, modern IC designs often employ double vias. Such protection requires additional resources (area), and must obey all design rules. These resources may be unavailable around some vias. In some congested areas, only a small subset of vias can be doubled [6.17]. Via doubling can be performed by modern commercial routers or by standalone yield enhancement tools after detailed routing.

Interconnect defects. The two most common manufacturing defects in wires are *shorts* (undesired connections) and *opens* (broken connections). To address shorts, adjacent wires can be spread further apart, which also decreases electromagnetic interference. However, spreading the wires too far can increase total wirelength, thereby increasing the design's exposure to opens. To address opens, *non-tree routing* [6.12] adds redundant wires to already routed nets. However, since increasing wirelength directly contradicts traditional routing objectives (Chaps. 5-6), this step is usually a post-processing step after detailed routing. Redundant wires increase the design's susceptibility to shorts, but make it immune to some opens.

Antenna-induced defects. Another type of manufacturing defect affects transistors, but can be mitigated by constraining routing topologies. It occurs after the transistor and one or more metal layers have been fabricated, but before other layers are completed. During plasma etching, metal wires not connected to PN-junction nodes may collect significant electric charges which, discharged through the gate dielectric (SiO_2 at older technology nodes, high-k dielectric at newer nodes), can irreversibly damage transistor gates. To prevent these *antenna effects*, detailed routers limit the ratio of metal to gate area on each metal layer. Specifically, they restrict the area of metal polygons connected to gates without being connected to a source/drain implant. When such *antenna rules* are violated, the simplest fix is to transfer a fraction of a route to a higher layer through a new or relocated via.

Some researchers have also proposed *manufacturability-aware routers*, where detailed routing explicitly optimizes yield. However, it is difficult to objectively quantify the benefit of such optimizations before manufacturing. As a result, such techniques have not yet caught on in the industry.

Chapter 6 Exercises

Exercise 1: Left-Edge Algorithm
Given a channel with the following pin connections (ordered left to right).
$TOP = [A\ B\ A\ 0\ E\ D\ 0\ F]$ and $BOT = [B\ C\ D\ A\ C\ F\ E\ 0]$.
(a) Find $S(col)$ for columns a-h and the minimum number of routing tracks.
(b) Draw the HCG and VCG.
(c) Use the left-edge algorithm to route this channel. For each track, mark the placed nets and draw the updated VCG from (b). Draw the channel with the fully routed nets.

Exercise 2: Dogleg Left-Edge Algorithm
Given a channel with the following pin connections (ordered left to right).
$TOP = [A\ A\ B\ 0\ A\ D\ C\ E]$ and $BOT = [0\ B\ C\ A\ C\ E\ D\ D]$.
(a) Draw the vertical constraint graph (VCG) without splitting the nets.
(b) Determine the zone representation for nets A-E. Find $S(col)$ for columns a-h.
(c) Draw the vertical constraint graph (VCG) with net splitting.
(d) Find the minimum number of required tracks with net splitting and without net splitting.
(e) Use the Dogleg left-edge algorithm to route this channel. For each track, state which nets are assigned. Draw the final routed channel.

Exercise 3: Switchbox Routing
Given the nets on each side of a switchbox,
(ordered bottom-to-top) $LEFT = [0\ G\ A\ F\ B\ 0]$ $RIGHT = [0\ D\ C\ E\ G\ 0]$
(ordered left-to-right) $BOT = [0\ A\ F\ G\ D\ 0]$ $TOP = [0\ A\ C\ E\ B\ D]$
Route the switchbox using the approach shown in the example in Sec. 6.4.2. For each column, mark the routed nets and their corresponding tracks. Draw the switchbox with all nets routed.

Exercise 4: Manufacturing Defects
Consider a region with high wiring congestion and a region where routes can be completed easily. For each type of manufacturing defect discussed in Sec. 6.6, is it more likely to occur in a congested region? Explain your answers. You may find it useful to visualize congested and uncongested regions using small examples.

Exercise 5: Modern Challenges in Detailed Routing
Develop an algorithmic approach to double-via insertion.

Exercise 6: Non-Tree Routing
Discuss advantages and drawbacks of non-tree routing (Sec. 6.6).

Chapter 6 References

[6.1]   C. J. Alpert, Z. Li, M. D. Moffitt, G.-J. Nam, J. A. Roy and G. Tellez, "What Makes a Design Difficult to Route", *Proc. Intl. Symp. on Phys. Design*, 2010, pp. 7-12.

[6.2]   D. Braun et al., "Techniques for Multilayer Channel Routing", *IEEE Trans. on CAD* 7(6) (1988), pp. 698-712.

[6.3]   J. P. Cohoon and P. L. Heck, "BEAVER: A Computational-Geometry-Based Tool for Switchbox Routing", *IEEE Trans. on CAD* 7(6) (1988), pp. 684-697.

[6.4]   J. Cong and C. L. Liu, "Over-the-Cell Channel Routing", *IEEE Trans. on CAD* 9(4) (1990), pp. 408-418.

[6.5]   J. Cong, D. F. Wong and C. L. Liu, "A New Approach to Three- or Four-Layer Channel Routing", *IEEE Trans. on CAD* 7(10) (1988), pp. 1094-1104.

[6.6]   D. N. Deutsch, "A 'Dogleg' Channel Router", *Proc. Design Autom. Conf.*, 1976, pp. 425-433.

[6.7]   S. H. Gerez and O. E. Herrmann, "Switchbox Routing by Stepwise Reshaping", *IEEE Trans. on CAD* 8(12) (1989), pp. 1350-1361.

[6.8]   A. Hashimoto and J. Stevens, "Wire Routing by Optimizing Channel Assignment within Large Apertures", *Proc. Design Autom. Workshop*, 1971, pp. 155-169.

[6.9]   T.-T. Ho, S. S. Iyengar and S.-Q. Zheng, "A General Greedy Channel Routing Algorithm", *IEEE Trans. on CAD* 10(2) (1991), pp. 204-211.

[6.10] N. D. Holmes, N. A. Sherwani and M. Sarrafzadeh, "Utilization of Vacant Terminals for Improved Over-the-Cell Channel Routing", *IEEE Trans. on CAD* 12(6) (1993), pp. 780-792.

[6.11] *International Technology Roadmap for Semiconductors*, 2009 edition, www.itrs.net.

[6.12] A. Kahng, B. Liu and I. Măndoiu, "Non-Tree Routing for Reliability and Yield Improvement", *Proc. Intl. Conf. on CAD*, 2002, pp. 260-266.

[6.13] J. Lienig, "Introduction to Electromigration-Aware Physical Design", *Proc. Intl. Symp. on Phys. Design*, 2006, pp. 39-46.

[6.14] W. K. Luk, "A Greedy Switchbox Router", *Integration, the VLSI J.* 3(2) (1985), pp. 129-149.

[6.15] J. K. Ousterhout et al., "Magic: A VLSI Layout System", *Proc. Design Autom. Conf.*, 1984, pp. 152-159.

[6.16] R. Rivest and C. Fiduccia, "A 'Greedy' Channel Router", *Proc. Design Autom. Conf.*, 1982, pp. 418-424.

[6.17] G. Xu, L.-D. Huang, D. Pan and M. Wong, "Redundant-Via Enhanced Maze Routing for Yield Improvement", *Proc. Asia and South Pacific Design Autom. Conf.*, 2005, pp. 1148-1151.

[6.18] T. Yoshimura, "An Efficient Channel Router", *Proc. Design Autom. Conf.*, 1984, pp. 38-44.

[6.19] T. Yoshimura and E. S. Kuh, "Efficient Algorithms for Channel Routing", *IEEE Trans. on CAD* 1(1) (1982), pp. 25-35.

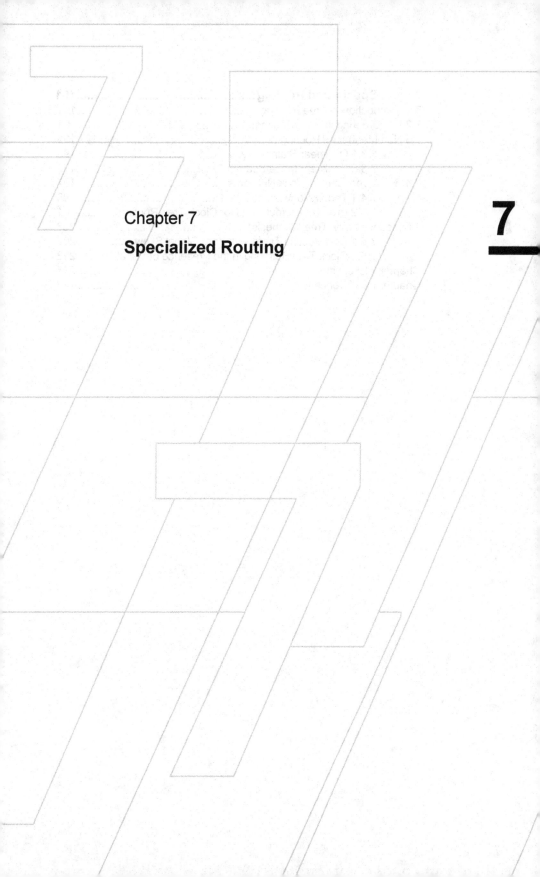

Chapter 7

Specialized Routing

7

7 Specialized Routing

For signal wires in digital integrated circuits, global routing (Chap. 5) is performed first, and detailed routing next (Chap. 6). However, some types of designs, such as analog circuits and printed circuit boards (PCBs) with gridless (trackless) routing, do not warrant this distinction. Smaller, older designs with only one or two metal layers also fall into this category. When global and detailed routing are not performed separately, *area routing* (Secs. 7.1-7.2) directly constructs metal routes for signal connections. Unlike routing with multiple metal layers, area routing emphasizes crossing minimization. Non-Manhattan routing is discussed in Sec. 7.3, and nets that require special treatment, such as clock signals, are discussed in Secs. 7.4-7.5.

7.1 Introduction to Area Routing

The goal of area routing is to route all nets in the design (1) without global routing, (2) within the given layout space, and (3) while meeting all geometric and electrical design rules. Area routing performs the following optimizations.

- minimizing the total routed length and number of vias of all nets
- minimizing the total area of wiring and the number of routing layers
- minimizing the circuit delay and ensuring an even wire density
- avoiding harmful capacitive coupling between neighboring routes

Area routing is performed subject to *technology* (number of routing layers, minimal wire width), *electrical* (signal integrity, coupling), and *geometry* (preferred routing directions, wire pitch) constraints. Electrical and technology constraints are traditionally represented by geometric rules, but modern routers seek to handle them directly to improve modeling accuracy. Nevertheless, reducing total wirelength may reduce circuit area, increase yield, and improve signal integrity. For example, the configuration on the left is preferred because its total length is minimal (Fig. 7.1).

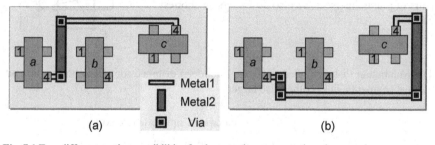

Fig. 7.1 Two different routing possibilities for the two-pin net connecting pins a_4 and c_4.

To measure wirelength, the (straight-line) *Euclidean* distance d_E and the (rectilinear) *Manhattan* distance d_M are used. For two points P_1 (x_1,y_1) and P_2 (x_2,y_2) in the plane, the Euclidean distance is defined as

$$d_E(P_1, P_2) = \sqrt{(x_2 - x_1)^2 + (y_2 - y_1)^2} = \sqrt{(\Delta x)^2 + (\Delta y)^2}$$

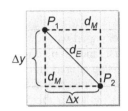

and the Manhattan distance is defined as

$$d_M(P_1, P_2) = |x_2 - x_1| + |y_2 - y_1| = |\Delta x| + |\Delta y|$$

By definition, Manhattan paths include only vertical and horizontal segments. Analog circuits and some MCMs can use unrestricted Euclidean routing while digital circuits use track-based Manhattan routing. The following facts and properties are relevant to VLSI routing.

Consider all shortest paths between points P_1 and P_2 in the plane. The Euclidean shortest path is unique, but there may be multiple Manhattan shortest paths. With no obstacles, the number of Manhattan shortest paths in an $\Delta x \times \Delta y$ region is

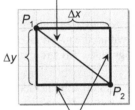

Euclidean Shortest Path

Manhattan Shortest Paths

$$\binom{\Delta x + \Delta y}{\Delta x} = \binom{\Delta x + \Delta y}{\Delta y} = \frac{(\Delta x + \Delta y)!}{\Delta x! \Delta y!}$$

The example (right) has 35 paths ($\Delta x = 4$ and $\Delta y = 3$).

Two pairs of points may admit two non-intersecting Manhattan shortest paths, while their Euclidean shortest paths intersect.

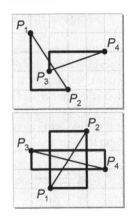

If all pairs of Manhattan shortest paths between two pairs of points intersect, then so do Euclidean shortest paths.

The Manhattan distance is equal to the Euclidean distance for single horizontal and vertical segments, but is otherwise larger.

$$\frac{d_M}{d_E} = \begin{cases} 1.41 & \text{worst case: a square where } \Delta x = \Delta y \\ 1.27 & \text{on average, without obstacles} \end{cases}$$

7.2 Net Ordering in Area Routing

The results and runtime of area routing can be very sensitive to the order in which nets are routed. This is especially true for Euclidean routing, which allows fewer shortest paths and is therefore prone to detours. Routing multiple nets by greedily optimizing the wirelength of one net at a time may produce inferior configurations with unnecessarily large numbers of routing failures (Fig. 7.2) and total wirelength (Fig. 7.3). Furthermore, multi-pin nets increase the complexity of net ordering when they are decomposed into two-pin subnets. Therefore, some algorithms determine net and pin ordering before routing.

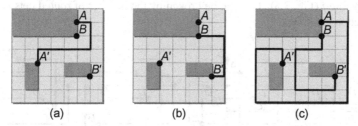

Fig. 7.2 Effect of net ordering on routability. (a) An optimal routing of net A prevents net B from being routed. (b) An optimal routing of net B prevents net A from being routed. (c) Nets A and B can be simultaneously routed only if each uses more than minimum wirelength.

The choice of net and pin ordering depends on the type of the routing algorithm used. Pin ordering can be optimized using, e.g., (1) Steiner tree-based algorithms (Sec. 5.6) or other methods to decompose multi-pin nets to two-pin nets, or (2) geometric criteria. For example, pin locations can be ordered by (non-decreasing) x-coordinate and connected from left to right. Given previously-connected pins, the next pin is connected using a shortest-path algorithm.

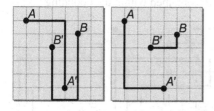

Fig. 7.3 Effect of net ordering on total wirelength, where routing net A first (left) is worse than routing net B first (right).

For n nets, there are $n!$ possible net orderings. In the absence of clear criteria and polynomial-time algorithms, constructive heuristics are used. These heuristics prioritize nets quantitatively or order them in pairs, as illustrated below. For a net *net*, let $MBB(net)$ be the minimum bounding box containing the pin locations of *net*, let $AR(net)$ be the aspect ratio of $MBB(net)$, and let $L(net)$ be the length of *net*.

Rule 1: For two nets i and j, if $AR(i) > AR(j)$, then i is routed before j (Fig. 7.4). Rationale: nets with square MBBs tend to have greater routing flexibility than nets with tall or wide bounding boxes (all straight nets have $AR = \infty$). If $AR(i) = AR(j)$, then ties can be broken by net length, i.e., if $L(i) < L(j)$, then i is routed before j.

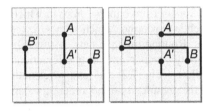

Fig. 7.4 Net ordering based on the aspect ratio of the nets' bounding boxes. Net A has higher aspect ratio ($AR(A) = \infty$) and results in shorter total wirelength (left), while routing net B first results in greater total wirelength.

Rule 2: For two nets i and j, if the pins of i are contained within $MBB(j)$, then i is routed before j (Fig. 7.5). Ties can be broken by the number of routed nets *not* fully contained in the MBB.

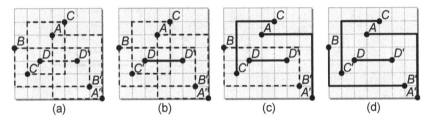

Fig. 7.5 Net ordering based on the pin locations inside the bounding boxes of the nets. The first-routed net has no pins contained within its MBB. Starting with net D, there are two potential net orderings: D-A-C-B and D-C-A-B. (a) Nets A-D with their MBBs. (b) Routing net D first. (c) Routing net C and then net A or net A and then net C. (d) Routing net B.

Rule 3: Let $\Pi(net)$ be the number of pins within $MBB(net)$ for net net. For two nets i and j, if $\Pi(i) < \Pi(j)$, then i is routed before j. That is, the net that is routed first has the smaller number of pins from other nets within its bounding box (Fig. 7.6). Ties are broken based on the number of pins that are contained within the bounding box.

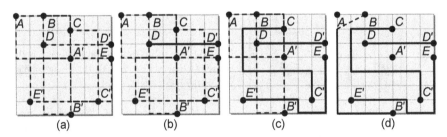

Fig. 7.6 Finding the net ordering based on the number of pins of nets within its MBB. (a) Nets A-E with their MBBs. (b) Net D is routed first because its MBB contains no pins. (c) Net C is routed because it contains one pin, and net E is routed next because it contains two pins. (d) Nets B and A are routed next, so the result is D-C-E-B-A. Note that this example cannot be routed with *sequential* net ordering on a Manhattan grid.

7.3 Non-Manhattan Routing   7.3

Recall from Sec. 7.1 that traditional Manhattan routing allows only vertical and horizontal segments. Shorter paths are possible with diagonal segments. However, arbitrary diagonal segments cannot be effectively manufactured. A possible compromise is to allow 45-degree or 60-degree segments in addition to horizontal and vertical segments. Such non-orthogonal routing configurations are commonly described by λ-geometry, where λ represents the number of possible routing directions[1] and the angles π / λ at which they can be oriented.

- $\lambda = 2$ (90 degrees): Manhattan routing (four routing directions)
- $\lambda = 3$ (60 degrees): *Y-routing* (six routing directions)
- $\lambda = 4$ (45 degrees): *X-routing* (eight routing directions)

The advantages of the latter two routing styles over Manhattan-based routing are decreased wirelength and via count. However, other steps in the physical design flow, such as physical verification, could take significantly longer. Additionally, non-Manhattan routing becomes prohibitively difficult at recent technology nodes due to limitations of optical lithography. Therefore, non-Manhattan routing is primarily employed on printed circuit boards (PCBs). This is illustrated by octilinear route planning in Sec. 7.3.1 and eight-directional path search in Sec. 7.3.2.

▶ **7.3.1 Octilinear Steiner Trees**

Octilinear Steiner minimum trees (*OSMT*) generalize rectilinear Steiner trees by allowing segments that extend in eight directions. The inclusion of diagonal segments gives more freedom when placing Steiner points, which may reduce total net length. Several OSMT algorithms have been proposed, such as in [7.9] and [7.19]. The following approach was developed by *Ho et al.* [7.9] (refer to the pseudocode on the next page).

First, find the shortest three-pin subnets of the net under consideration. To identify these three-pin groups, the *Delaunay triangulation*[2] is found over all pins (line 2). Second, sort all the groups in ascending order of their minimum octilinear routed lengths (line 3). Then, integrate these three-pin subnets into the overall OSMT. For each group *subT* in sorted order (line 4), (1) route *subT* with the minimum octilinear length (line 5), (2) merge *subT* with the current octilinear Steiner tree *OST* (line 6), and (3) locally optimize *OST* based on *subT* (line 7).

---

[1] Not to be confused with the layout-scaling parameter λ.

[2] The Delaunay triangulation for a set of points P in a plane is a triangulation $DT(P)$ such that no points in P lie inside the circumcircle of any triangle in $DT(P)$. The circumcircle of a triangle *tri* is defined as a circle which passes through all the vertices of *tri*.

Octilinear Steiner Tree Algorithm [7.9]

Input: set of all pins P and their coordinates

Output: heuristic octilinear minimum Steiner tree OST

1. $OST = \emptyset$
2. T = set of all three-pin nets of P found by Delaunay triangulation
3. $sortedT$ = SORT(T,minimum octilinear distance)
4. **for** (i = 1 **to** $|sortedT|$)
5. $subT$ = ROUTE($sortedT[i]$)   // route minimum tree over $subT$
6.   ADD($OST,subT$)   // add route to existing tree
7.   IMPROVE($OST,subT$)   // locally improve OST based on $subT$

Example: Octilinear Steiner Tree

Given: pins P_1-P_{12}.

   P_1 (2,17) P_2 (1,14) P_3 (11,15) P_4 (4,11)
   P_5 (14,12) P_6 (2,9) P_7 (11,9) P_8 (12,6)
   P_9 (16,6) P_{10} (7,4) P_{11} (3,1) P_{12} (14,1)

Task: find a heuristic octilinear Steiner minimum tree.

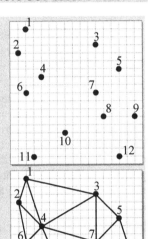

Solution:

Find all three-pin subnets using Delaunay triangulation. Find the minimum octilinear Steiner tree cost for each subnet and sort the subnets in ascending order of this cost.

$L(T_{2,4,6}) \approx 7.0$	$L(T_{7,8,9}) \approx 7.4$	$L(T_{1,2,4}) \approx 7.6$
$L(T_{3,5,7}) \approx 8.4$	$L(T_{8,9,12}) \approx 8.6$	$L(T_{7,8,10}) \approx 8.6$
$L(T_{4,6,10}) \approx 9.6$	$L(T_{5,7,9}) \approx 9.6$	$L(T_{8,10,12}) \approx 10.6$
$L(T_{6,10,11}) \approx 11.8$	$L(T_{3,4,7}) \approx 12.6$	$L(T_{4,7,10}) \approx 12.6$
$L(T_{10,11,12}) \approx 13.4$	$L(T_{1,3,4}) \approx 13.8$	

Add $T_{2,4,6}$ to OST. No optimization is necessary because it is the first merged subtree. This is also the case for trees $T_{7,8,9}$, $T_{1,2,4}$, and $T_{3,5,7}$. Merging $T_{8,9,12}$ causes a cycle between the new Steiner point in $T_{8,9,12}$, P_8 and P_9.

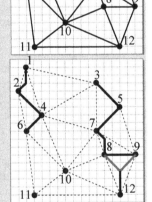

To resolve this conflict, find the minimum octilinear distance needed to connect the three new points. The minimal tree uses the two diagonal segments, $2\sqrt{2^2 + 2^2} \approx 5.7$, which results in smaller wirelength than the horizontal segment connecting P_8 and P_9 plus a vertical segment, $4 + 2 = 6$.

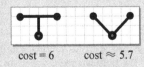

cost = 6   cost ≈ 5.7

Continue merging the remaining subtrees. The final heuristic octilinear minimum Steiner tree is constructed after merging all subtrees.

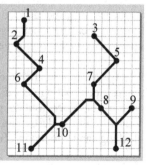

▶ 7.3.2 Octilinear Maze Search

Route planning based on OSMTs assumes octilinear detailed routing. One approach is based on Lee's wave-propagation algorithm [7.13] where instead of only expanding nodes in four directions, eight directions are used, including diagonals. The wave propagation begins at s and marks all previously-unvisited neighbors with index 1 (Fig. 7.7(a)). From each node with index 1, the propagation expands again, and each previously-unvisited neighbor is marked with index 2 (Fig. 7.7(b)). Such iterations continue until the wave reaches the target t or no further propagation is possible. After t is reached, a path is backtraced by stepping to next smallest index until s is reached (Fig. 7.7(c)).

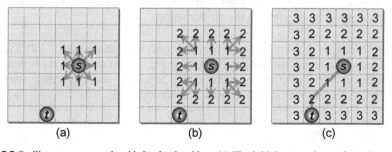

Fig. 7.7 Octilinear maze search with Lee's algorithm. (a) The initial outward pass from the source node s. (b) The second pass, branching out in all eight directions. (c) The third pass, where the target node t has been found, and a path is traced back from t to s.

7.4 Basic Concepts in Clock Networks

Most digital designs are *synchronous* – computation progresses when current values of internal state variables and input variables are fed to combinational logic networks, which then generate outputs as well as the next values of the state variables. A *clock signal* or "heartbeat", required to maintain synchronization of all computation that takes place across the chip, may be generated off-chip, or by special analog circuitry such as *phase-locked loops* or *delay-locked loops*. Its frequency may be divided or multiplied depending on the needs of individual blocks. Once the clock signal entry points and *sinks* (flip-flops and latches) are known, *clock tree routing* generates a clock tree for each clock domain of the circuit.

The special role of the clock signal in synchronizing all computations on the chip makes clock routing very different from the other types of routing (Chaps. 5-6). The crux of the clock routing problem is that the signal must be delivered from the source to all the destinations, or sinks, at the same time. Advanced algorithms for clocking can be found in Chaps. 42-43 of [7.1] and in books [7.15] and [7.18].

► **7.4.1 Terminology**

A *clock routing problem instance* (*clock net*) is represented by $n + 1$ terminals, where s_0 is designated as the *source*, and $S = \{s_1, s_2, \dots, s_n\}$ is designated as *sinks*. Let s_i, $0 \le i \le n$, denote both a terminal and its location.

A *clock routing solution* consists of a set of wire segments that connect all of the terminals of the clock net, so that a signal generated at the source can be propagated to all of the sinks. The clock routing solution has two aspects – its *topology* and its *embedding*.

The *clock tree topology* (*clock tree*) is a rooted binary tree G with n leaves corresponding to the set of sinks. Internal nodes of the topology correspond to the source and any Steiner points in the clock routing.

The *embedding* of a given clock tree topology provides exact physical locations of the edges and internal nodes of the topology. Fig. 7.8(a) shows a six-sink clock tree instance. Fig. 7.8(b) shows a connection topology, and Fig. 7.8(c) shows a possible clock tree solution that embeds that topology.

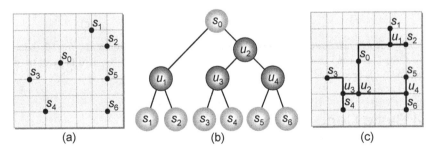

Fig. 7.8 (a) A clock tree routing instance. (b) Connection topology. (c) Embedding.

Assume that the clock tree T is oriented such that the source terminal s_0 is the *root* of the tree. Each node $v \in T$ is connected to its parent $p \in T$ by edge (p,v). The cost of each edge $e \in T$ is its wirelength, denoted by $|e|$. Let the cost of T, denoted by $cost(T)$, be the sum of its (embedded) edge costs.

Signal delay is the time required for a signal transition (low to high, or high to low) to propagate from one node to another node, e.g., in a routing tree. Signal transitions are initiated at the outputs of logic gates, which are constructed from transistors that have highly nonlinear behavior. The transitions propagate through complex wire and via structures that have parasitic resistances, capacitances and inductances. Hence, it is difficult to exactly calculate signal delay. Circuit simulators such as *SPICE*, or commercial timing analysis tools such as *PrimeTime*, are used to obtain accurate "signoff delay" calculations during the final checking steps before the design is sent to production. However, to guide place-and-route algorithms, considerably less accuracy is needed. Two common signal delay estimates used in timing-driven routing are the *linear* and *Elmore* delay models. The following is a reproduction of the development given in [7.11].

In the *linear delay* model, signal delay from s_i to s_j is proportional to the length of the $s_i \sim s_j$ path in the routing tree and is independent of the rest of the connection topology. Thus, the normalized linear delay between any two nodes u and w in a source-sink path is the sum of the edge lengths $|e|$ in the $u \sim w$ path

$$t_{LD}(u, w) = \sum_{e \in (u \sim w)} |e|$$

On-chip wires are passive, *resistive-capacitive* (*RC*) structures, for which both resistance (*R*) and capacitance (*C*) typically grow in proportion to the length of the wire. Thus, the linear delay model does not accurately capture the "quadratically growing" *RC* component of wire delay. On the other hand, the linear approximation provides reasonable guidance to design tools, especially for older technologies that have smaller drive resistance of transistors and larger wire widths (smaller wire resistances). In practice, the linear delay model is very convenient to use in EDA software tools because of its ease of evaluation.

In the *Elmore delay* model, given the routing tree T with root (source) node s_0,

- (p,v) denotes the edge connecting node v to its parent node p in T
- $R(e)$ and $C(e)$ denote the respective resistance and capacitance of edge $e \in T$
- T_v denotes the subtree of T rooted at v
- $C(v)$ denotes the sink capacitance of v
- $C(T_v)$ denotes the *tree capacitance* of T_v, i.e., the sum of sink and edge capacitances in T_v

If node v is a terminal, then $C(v)$ is typically the capacitance of the input pin to which the clock signal is routed. If node v is a Steiner node, then $C(v) = 0$. If T_v is a single (leaf) node, $C(T_v)$ is equal to v's sink capacitance $C(v)$.

Using this notation, the Elmore delay approximation for an edge (p,v) is

$$t_{ED}(p,v) = R(p,v) \cdot \left(\frac{C(p,v)}{2} + C(v) \right)$$

This can be seen as a sum of RC delay products, with the factor of one-half corresponding to a ~63% threshold delay. Last, if R_d denotes the on-resistance of the output transistor at the source ("stronger" driving gates will have smaller on-resistance values R_d), then the Elmore delay $t_{ED}(s)$ for sink s

$$t_{ED}(s) = R_d C(s_0) + \sum_{e \in (s_0, s)} t_{ED}(e)$$

Physical design tools use the Elmore delay approximation for three main reasons. First, it accounts for the sink delay impact of off-path wire capacitance – the edges of the routing tree that are not directly on the source-to-sink path. Second, it offers reasonable accuracy and good *fidelity* (correlation) with respect to accurate delay estimates from circuit simulators. Third, it can be evaluated at all nodes of a tree in time that is *linear* in tree size (number of edges). This is realized by two *depth-first traversals*: the first calculates the tree capacitance $C(T_v)$ below each node in the tree, while the second calculates the delays from the source to each node [7.11].

Clock *skew* is the (maximum) difference in clock signal arrival times between sinks. This parameter of the clock tree solution is important, since the clock signal must be delivered to all sinks at the same time. If $t(u,v)$ denotes the signal delay between nodes u and v, then the *skew* of clock tree T is

$$skew(T) = \max_{s_i, s_j \in S} |t(s_0, s_i) - t(s_0, s_j)|$$

If there exists a path of combinational logic from the (data) output pin of one sink to the (data) input pin of another sink, then the two sinks are said to be *related* or *sequentially adjacent*. Otherwise, the two sinks are *unrelated*.

Local skew is the maximum difference in arrival times of the clock signal at the clock pins of two or more related sinks.

Global skew is the maximum difference in arrival times of the clock signal at the clock pins of any two (related or unrelated) sinks – i.e., the difference between shortest and longest source-sink path delays in the clock distribution network. In practice, skew typically refers to global skew.

▶ 7.4.2 Problem Formulations for Clock-Tree Routing

This section presents some basic clock routing formulations. The most fundamental is the *zero-skew tree* problem. Practical variations include the *bounded-skew tree* and *useful-skew tree* problems. The integration of zero-skew trees in a modern, low-power clock-network design flow is further discussed in Sec. 7.5.2, with more details in [7.14]. It relies on SPICE – software for circuit simulation – for a high degree of accuracy.

Zero skew. If a clock tree exhibits zero skew, then it is a *zero-skew tree* (*ZST*). For skew to be well-defined, a delay estimate (e.g., linear or Elmore delay) is implicit.

Zero-Skew Tree (ZST) Problem. Given a set S of sink locations, construct a ZST $T(S)$ with minimum cost. In some contexts, a connection topology G is also given.

Bounded skew. While the ZST problem leads to elegant physical design algorithms that form the basis of commercial solutions, practical clock tree routing does not typically achieve exact zero skew.

In practice, a "true ZST" is not desirable. ZSTs can use a significant amount of wirelength, increasing the total capacitance of the network. Moreover, a true ZST is also not achievable in practice – manufacturing variability for both transistors and interconnects can cause differences in the RC constants of wire segments of a given layer. Thus, signoff timing analysis is with respect to a non-zero *skew bound* that must be achieved by the clock routing tool.

Bounded-Skew Tree (BST) Problem. Given a set S of sink locations and a skew bound $UB > 0$, construct a clock tree $T(S)$ with $skew(T(S)) \leq UB$ having minimum cost. As with the ZST problem, in certain contexts a topology G may be specified. Notice that when the skew is unbounded ($UB = \infty$), the BST problem becomes the classic RSMT problem (Chap. 5).

Useful skew. Clock trees do not always require bounded *global* skew. Correct chip timing only requires control of the *local* skews between pairs of related flip-flops or latches. While the clock tree routing problem can be conveniently formulated in terms of global skew, this actually over-constrains the problem. The increasingly prominent *useful skew* formulation is based on analysis of local skew constraints.

In synchronous circuits, the data signal that propagates from a flip-flop (sink) output to the next flip-flop input should arrive neither too late nor too early. The former failure mode (late arrival) is *zero clocking*, while the latter (early arrival) is *double clocking* [7.7]. In contrast to formulations that minimize or bound global skew, *Fishburn* [7.7] proposed a clock skew optimization method that introduces *useful skew* – perturbing clock arrival times at sinks – in the clock tree to either minimize the clock period or maximize the clock *safety* margin. The clock period P can be reduced by appropriate choices of sink arrival times (Fig. 7.9).

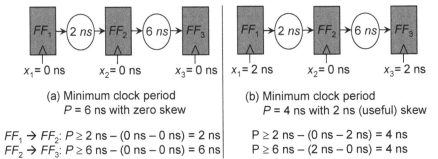

(a) Minimum clock period
P = 6 ns with zero skew

(b) Minimum clock period
P = 4 ns with 2 ns (useful) skew

$FF_1 \rightarrow FF_2$: $P \geq 2$ ns – (0 ns – 0 ns) = 2 ns	$P \geq 2$ ns – (0 ns – 2 ns) = 4 ns
$FF_2 \rightarrow FF_3$: $P \geq 6$ ns – (0 ns – 0 ns) = 6 ns	$P \geq 6$ ns – (2 ns – 0 ns) = 4 ns

Fig. 7.9 Example of useful skew for clock cycle time reduction. (a) Zero skew results in a 6 ns clock period. (b) Useful skews of 2 ns, 0 ns and 2 ns at x_1, x_2 and x_3 result in a 4 ns clock period.

To avoid zero clocking, the data edge generated by FF_i due to a clock edge must arrive at FF_j no later than t_{setup} before the earliest arrival of the next clock edge. Formally, $x_i + t_{setup} + \max(i,j) \leq x_j + P$ must be met with clock period P, where

– x_i is the latest time at which the clock edge can arrive at FF_i
– $\max(i,j)$ is the slowest (longest) signal propagation from FF_i to FF_j
– $x_j + P$ is the earliest arrival time of the next clock edge at FF_j

To avoid double clocking between two flip-flops FF_i and FF_j, the data edge generated at FF_i due to a clock edge must arrive at FF_j no sooner than t_{hold} after the latest possible arrival of the same clock edge. Formally, $x_i + \min(i,j) \geq x_j + t_{hold}$ must be met, where

– x_i is the earliest time at which the clock edge can arrive at FF_i
– $\min(i,j)$ denote the fastest (shortest) signal propagation from FF_i to FF_j
– x_j be the latest arrival time of the clock at FF_j

Fishburn observed that linear programming can be used to find optimal clock arrival times x_i at all sinks to either (1) minimize clock period (*LP_SPEED*), or (2) maximize the safety margin (*LP_SAFETY*).

Useful Skew Problem (LP_SPEED). Given (1) constant values of t_{setup} and t_{hold}, (2) maximum and minimum signal propagation times $\max(i,j)$ and $\min(i,j)$ between all pairs (i,j) of related sinks, and (3) minimum source-sink delay t_{min}, determine clock arrival times x_i for all sinks to minimize *clock period P*, subject to the following constraints.

$$x_i - x_j \geq t_{hold} - \min(i,j) \qquad \text{for all related } (i,j)$$
$$x_j - x_i + P \geq t_{setup} + \max(i,j) \qquad \text{for all related } (i,j)$$
$$x_i \geq t_{min} \qquad \text{for all } i$$

Useful Skew Problem (*LP_SAFETY*). Given (1) constant values of t_{setup} and t_{hold}, (2) maximum and minimum signal propagation times $max(i,j)$ and $min(i,j)$ between all pairs (i,j) of related sinks, and (3) minimum source-sink delay t_{min}, determine clock arrival times x_i for all sinks to maximize *safety margin SM*, subject to

$$x_i - x_j - SM \geq t_{hold} - min(i,j) \qquad \text{for all related } (i,j)$$
$$x_j - x_i - SM \geq t_{setup} + max(i,j) - P \qquad \text{for all related } (i,j)$$
$$x_i \geq t_{min} \qquad \text{for all } i$$

7.5 Modern Clock Tree Synthesis   7.5

Clock trees play a vital role in modern synchronous designs and significantly impact the circuit's performance and power consumption. A clock tree should have low skew, simultaneously delivering the same signal to every sequential gate. After the initial tree construction (Sec. 7.5.1), the clock tree undergoes clock buffer insertion and several subsequent skew optimizations (Sec. 7.5.2).

► **7.5.1 Constructing Trees with Zero Global Skew**

This section presents five early algorithms for clock tree construction whose underlying concepts are still used in today's commercial EDA tools. Several scenarios are covered, including algorithms that (1) construct a clock tree independent of the clock sink locations, (2) construct the clock tree topology and embedding simultaneously, and (3) construct only the embedding given a clock tree topology as input.

H-tree. The H-tree is a self-similar, fractal structure (Fig. 7.10) with exact zero skew due to its symmetry. It was first popularized by *Bakoglu* [7.2]. In the unit square, a segment is passed through the root node at center, then two shorter line segments are constructed at right angles to the first segment, to the centers of the four quadrants; this process continues recursively until the sinks are reached. The H-tree is frequently used for top-level clock distribution, but cannot be employed directly for the entire clock tree due to (1) blockages, (2) irregularly placed clock sinks, and (3) excessive routing cost. That is, to reach all $n = 4^k$ sinks uniformly located in the unit square, where $k \geq 1$ is the number of levels in the H-tree, the wirelength of the H-tree grows as $3\sqrt{n}/2$. To minimize signal reflections at branching points, the wire segments can be *tapered* – halving the wire width at each branching point encountered as one moves away from the source.

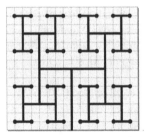

Fig. 7.10 An H-tree.

Method of Means and Medians (MMM). The method of means and medians (MMM) was proposed by *Jackson, Srinivasan* and *Kuh* in 1990 [7.10] to overcome the topology limitations of the H-tree. MMM is applicable even when the clock terminals are arbitrarily arranged. The basic idea is to recursively partition the set of terminals into two subsets of equal cardinality (median). Then, the center of mass of the set is connected to the centers of mass of the two subsets (mean) (Fig. 7.11). The basic MMM algorithm is described below.

Basic Method of Means and Medians (BASIC_MMM(S,T))
Input: set of sinks S, empty tree T
Output: clock tree T

1.	**if** ($|S| \leq 1$)	
2.	**return**	
3.	$(x_0,y_0) = (x_c(S),y_c(S))$	// center of mass for S
4.	$(S_A,S_B) = $ PARTITION(S)	// median to determine S_A and S_B
5.	$(x_A,y_A) = (x_c(S_A),y_c(S_A))$	// center of mass for S_A
6.	$(x_B,y_B) = (x_c(S_B),y_c(S_B))$	// center of mass for S_B
7.	ROUTE(T,x_0,y_0,x_A,y_A)	// connect center of mass of S to
8.	ROUTE(T,x_0,y_0,x_B,y_B)	// center of mass of S_A and S_B
9.	BASIC_MMM(S_A,T)	// recursively route S_A
10.	BASIC_MMM(S_B,T)	// recursively route S_B

Let $(x_c(S),y_c(S))$ denote the x- and y- coordinates for the center of mass of the point set S, defined as

$$x_c(S) = \frac{\sum_{i=1}^{n} x_i}{n} \quad \text{and} \quad y_c(S) = \frac{\sum_{i=1}^{n} y_i}{n}$$

S_A and S_B are the two equal-cardinality subsets obtained when S is partitioned by a median.

While the MMM strategy may be simplified to a *top-down* H-tree construction, the clock skew is minimized only heuristically. The maximum difference between two source-sink pathlengths can be very large (up to diameter of the chip) in the worst case. Thus, the algorithm's effectiveness depends heavily on the choice of cut directions for median computation.

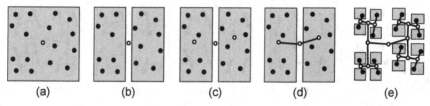

Fig. 7.11 Illustration of the main steps of the method of means and medians (MMM). (a) Find the center of mass (white 'o') of the point set S (black 'o'). (b) Partition S by the median. (c) Find the center of mass for the left and right subsets of S. (d) Route (connect) the center of mass of S with the centers of mass of the left and right subsets of S. (e) Final result after recursively performing MMM on each subset.

Recursive Geometric Matching (RGM). The recursive geometric matching (RGM) algorithm [7.12] was proposed in 1991. Whereas MMM is a top-down algorithm, RGM proceeds in a bottom-up fashion. The basic idea is to recursively find a minimum-cost geometric matching of n sinks – a set of $n / 2$ line segments that connect n endpoints pairwise, with no two line segments sharing an endpoint and minimal total segment length. After each matching step, a *balance* or *tapping* point is found on each matching segment to preserve zero skew to the associated sinks. The set of $n / 2$ tapping points then forms the input to the next matching step. The algorithm is illustrated in Fig. 7.12.

Formally, let T denote a rooted binary tree, let S denote a set of points, and let M denote a set of matching pairs $<P_i,P_j>$ over S. The *clock entry point* (*CEP*) denotes the root of the clock tree, i.e., the location from which the clock signal is propagated.

Recursive Geometric Matching Algorithm
Input: set of sinks S, empty tree T
Output: clock tree T
1.　**if** ($|S| \leq 1$)
2.　　**return**
3.　M = min-cost geometric matching over S
4.　$S' = \varnothing$
5.　**foreach** ($<P_i,P_j> \in M$)
6.　　T_{P_i} = subtree of T rooted at P_i
7.　　T_{P_j} = subtree of T rooted at P_j
8.　　tp = tapping point on (P_i,P_j)　　　// point that minimizes the skew of
　　　　　　　　　　　　　　　　　　　　// 　the tree $T_{tp} = T_{P_i} \cup T_{P_j} \cup (P_i,P_j)$
9.　　ADD(S',tp)　　　　　　　　　　　// add tp to S'
10.　ADD($T,(P_i,P_j)$)　　　　　　　　　// add matching segment (P_i,P_j) to T
11.　**if** ($|S| \% 2 == 1$)　　　　　　　　// if $|S|$ is odd, add unmatched node
12.　　ADD(S', unmatched node)
13.　RGM(S',T)　　　　　　　　　　　// recursively call RGM

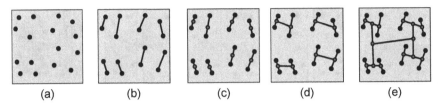

Fig. 7.12 Illustration of the recursive geometric matching (RGM) algorithm. (a) Set of n sinks S. (b) Min-cost geometric matching on n sinks. (c) Find balance or tapping points (gray 'o') for each $n / 2$ line segments. Notice that each tapping point is not necessarily the midpoint of the matching segment; it is the point that achieves zero skew in the subtree. (d) Min-cost geometric matching on $n - 2$ tapping points. (e) Final result after recursively performing RGM on each newly generated set of tapping points.

In practice, RGM improves clock tree balance and wirelength over MMM. Fig. 7.13 compares the results from (a) MMM and (b) RGM on a set of four sinks. For this instance, when MMM makes an unfortunate choice of cut direction (median computation), RGM can reduce wirelength by a factor of two. However, as with the MMM algorithm, RGM does not guarantee zero skew. In particular, if two subtrees have very different source-sink delays and their roots are matched, then it may not be possible to find a zero-skew tapping point on the matching segment.

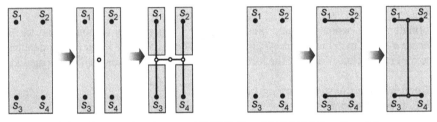

Fig. 7.13 Wirelength comparison between MMM (left) and RGM (right) on four sinks.

Exact Zero Skew. The exact zero skew algorithm, proposed in 1991 [7.17], adopts a bottom-up process of matching subtree roots and merging the corresponding subtrees, similar to RGM. However, this algorithm features two important improvements. First, it finds exact zero-skew tapping points with respect to the Elmore delay model rather than the linear delay model. This makes the result more useful, with smaller actual clock skew in practice. Second, it maintains exact delay balance even when two subtrees with very different source-sink delays are matched. This is accomplished by elongating wires as necessary to equalize source-sink delays.

When the roots of two subtrees are matched and the subtrees are merged, a zero-skew merging (tapping) point is determined (Fig. 7.14). In the figure, tp indicates the position of the zero-skew tapping point along the matching segment of length $L(s_1, s_2)$ between nodes s_1 and s_2.

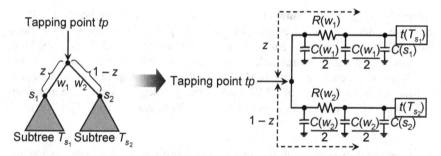

Fig. 7.14 Finding a zero-skew tapping point when two subtrees are merged. The tapping point *tp*, located on the segment connecting both subtrees, is where the Elmore delay to sinks is equalized.

To achieve zero skew, the delays from the tapping point *tp* to the sinks of T_{s_1}, and from the tapping point *tp* to the sinks of T_{s_2}, must be the same. Thus,

$$t_{ED}(tp) = R(w_1) \cdot \left(\frac{C(w_1)}{2} + C(s_1) \right) + t_{ED}(T_{s_1}) = R(w_2) \cdot \left(\frac{C(w_2)}{2} + C(s_2) \right) + t_{ED}(T_{s_2})$$

where w_1 is the wire segment connecting s_1 to *tp*, w_2 is the wire segment connecting *tp* to s_2, $R(w_1)$ and $C(w_1)$ denote the respective resistance and capacitance of w_1, $R(w_2)$ and $C(w_2)$ denote the respective resistance and capacitance of w_2, $C(s_1)$ is the capacitance of s_1, $C(s_2)$ is the capacitance of s_2, $t_{ED}(T_{s_1})$ is the Elmore delay of subtree T_{s_1}, and $t_{ED}(T_{s_2})$ is the Elmore delay of subtree T_{s_2}.

The wire resistance and capacitance of w_1 are found by multiplying the respective unit resistance α and capacitance β to the distance from s_1 to *tp*.

$$R(w_1) = \alpha \cdot z \cdot L(s_1,s_2) \text{ and } C(w_1) = \beta \cdot z \cdot L(s_1,s_2)$$

Similarly, the wire resistance and capacitance of w_2 are found by multiplying the respective unit resistance α and capacitance β with the distance from *tp* to s_2.

$$R(w_2) = \alpha \cdot (1-z) \cdot L(s_1,s_2) \text{ and } C(w_2) = \beta \cdot (1-z) \cdot L(s_1,s_2)$$

Combining the above equations yields the position of the zero-skew tapping point

$$z = \frac{(t(T_{s_2}) - t(T_{s_1})) + \alpha \cdot L(s_1,s_2) \cdot \left(C(s_2) + \dfrac{\beta \cdot L(s_1,s_2)}{2} \right)}{\alpha \cdot L(s_1,s_2) \cdot (\beta \cdot L(s_1,s_2) + C(s_1) + C(s_2))}$$

When $0 \le z \le 1$, the tapping point is located along the segment connecting the roots of the two subtrees. Otherwise, the wire must be elongated to meet the zero-skew condition.

Deferred-Merge Embedding (DME). The deferred-merge embedding (DME) algorithm defers the choice of merging (tapping) points for subtrees of the clock tree. DME *optimally* embeds any given topology over the sink set S: the embedding has minimum possible source-sink linear delay, and minimum possible total tree cost. While the preceding methods MMM and RGM each require only a set of sink locations as input, variants of DME needs a tree topology as input. The algorithm was independently proposed by several groups – *Boese* and *Kahng* [7.3], *Chao et al.* [7.4], and *Edahiro* [7.6].

A fundamental weakness of the preceding algorithms is that they decide the locations of internal nodes of the clock tree very early – before intelligent decisions are even possible. Once a centroid is determined in MMM, or a tapping point in RGM or an exact zero skew tree, it is never changed. Yet, in the Manhattan geometry, two sinks in general position will have an *infinite* number of midpoints, creating a tilted line segment, or *Manhattan arc* (Fig. 7.15); each of these midpoints affords the same minimum wirelength and exact zero skew. Ideally, the selection of embedding points for internal nodes will be delayed for as long as possible.

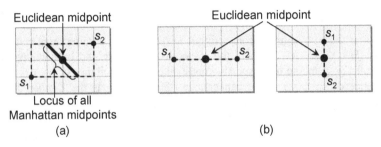

Fig. 7.15 The locus of all midpoints between two sinks s_1 and s_2 is a *Manhattan arc* in the Manhattan geometry. On the other hand, the midpoint is unique in Euclidean geometry. (a) Sinks s_1 and s_2 are not horizontally aligned. Therefore, the Manhattan arc has non-zero length. (b) Sinks s_1 and s_2 are horizontally (left) and vertically (right) aligned. Therefore, the Manhattan arc for both cases has zero length.

The DME algorithm embeds internal nodes of the given topology G via a two-phase process. The first phase of DME is bottom-up, and determines all possible locations of internal nodes of G that are consistent with a minimum-cost ZST T. The output of the first phase is a "tree of line segments", with each line segment being the locus of possible placements of an internal node of T. The second phase of DME is top-down, and chooses the exact locations of all internal nodes in T. The output of the second phase is a fully embedded, minimum-cost ZST with topology G.

In the following, let S denote the set of sinks $\{s_1, s_2, \dots, s_n\}$, and let s_0 denote the clock source. Let $pl(v)$ denote the location of a node v in the output clock tree T. Several terms are specific to the Manhattan geometry. As already noted, a *Manhattan arc* is a tilted line segment with slope ± 1. If the two sinks of Fig. 7.15 are horizontally or vertically aligned, there is only one possible midpoint, i.e., a zero-length Manhattan arc.

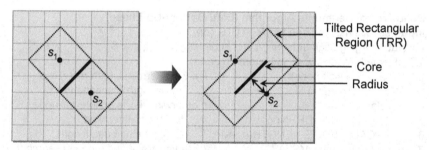

Fig. 7.16 (a) Sinks s_1 and s_2 form a Manhattan arc. (b) An example of a tilted rectangular region (TRR) for the Manhattan arc of s_1 and s_2 with radius of two units.

A *tilted rectangular region* (*TRR*) is a collection of points within a fixed distance of a Manhattan arc (Fig. 7.16). The *core* of a TRR is the subset of its points at maximum distance from its boundary, and the *radius* of a TRR is the distance between its core and boundary.

The *merging segment* of a node v in the topology, denoted by $ms(v)$, is the locus of feasible locations for v, consistent with exact zero skew and minimum wirelength (Fig. 7.17). The following presents the sub-algorithms used for DME.

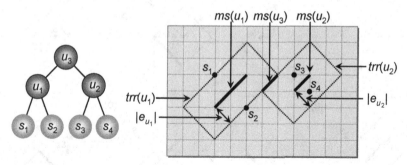

Fig. 7.17 A bottom-up construction of the merging segment $ms(u_3)$ for node u_3, the parent of nodes u_1 and u_2, given the topology on the left. The sinks s_1 and s_2 form the merging segment $ms(u_1)$, and the sinks s_3 and s_4 form the merging segment $ms(u_2)$. The two segments $ms(u_1)$ and $ms(u_2)$ together form the merging segment $ms(u_3)$.

The bottom-up phase of DME (Build Tree of Segments algorithm) starts with all sink locations S given. Each sink location is viewed as a (zero-length) Manhattan arc (lines 2-3). If two sinks have the same parent node u, then the locus of possible placements of u is a *merging segment* (Manhattan arc) $ms(u)$. In general, given the Manhattan arcs that are the merging segments of two nodes a and b, the merging segment of their parent node is uniquely determined due to the minimum-cost property, and is itself another Manhattan arc (Fig. 7.17). The edge lengths $|e_a|$ and $|e_b|$ are uniquely determined by the minimum-length and zero-skew requirements (lines 5-11). As a result, the entire tree of merging segments can be constructed bottom-up in linear time (Fig. 7.18).

Build Tree of Segments Algorithm (DME Bottom-Up Phase)
Input: set of sinks and tree topology $G(S, Top)$
Output: merging segments $ms(v)$ and edge lengths $|e_v|$, $v \in G$

1. **foreach** (node $v \in G$, in bottom-up order)
2.   **if** (v is a sink node)   // if v is a terminal, then $ms(v)$ is a
3. $ms[v] = PL(v)$ //   zero-length Manhattan arc
4.   **else**   // otherwise, if v is an internal node,
5. (a,b) = CHILDREN(v)   //   find v's children and
6.   CALC_EDGE_LENGTH(e_a, e_b) //   calculate the edge length
7. $trr[a][core] = MS(a)$ // create $trr(a)$ – find merging segment
8. $trr[a][radius] = |e_a|$ //   and radius of a
9. $trr[b][core] = MS(b)$ // create $trr(b)$ – find merging segment
10. $trr[b][radius] = |e_b|$ //   and radius of b
11. $ms[v] = trr[a] \cap trr[b]$ // merging segment of v

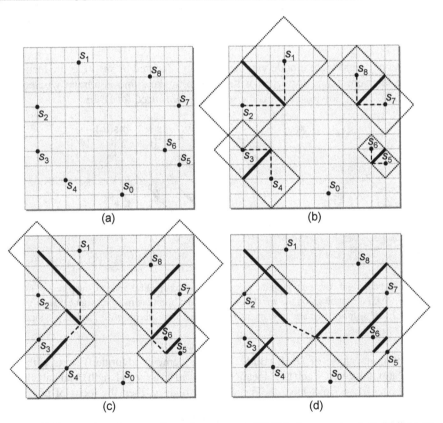

(a)   (b)

(c)   (d)

Fig. 7.18 Construction of a tree of merging segments (DME bottom-up phase). Solid lines are merging segments, dotted rectangles are the tilted rectangular regions (TRR), and dashed lines are edges between merging segments; s_0 is the clock source, and s_1-s_8 are sinks. (a) The eight sinks and the clock source. (b) Construct merging segments for the eight sinks. (c) Construct merging segments for the segments generated in (a). (d) Construct the root segment, the merge segment that connects to the clock source.

Find Exact Locations (DME Top-Down Phase)
Input: set of sinks *S*, tree topology *G*, outputs of DME bottom-up phase
Output: minimum-cost zero-skew tree *T* with topology *G*
1.   **foreach** (non-sink node $v \in G$ in top-down order)
2.   **if** (*v* is the root)
3.   *loc* = any point in *ms(v)*
4.   **else**
5.   *par* = PARENT(*v*)   // *par* is the parent of *v*
6.   *trr[par][core]* = *PL(par)*   // create *trr(par)* – find merging segment
7.   *trr[par][radius]* = $|e_v|$ //   and radius of *par*
8.   *loc* = any *point* in *ms[v]* $\cap$ *trr[par]*
9.   *pl[v]* = *loc*

In the DME top-down phase (Find Exact Locations algorithm), exact locations of internal nodes in *G* are determined, starting with the root. Any point on the root merging segment from the bottom-up phase is consistent with a minimum-cost ZST (lines 2-3). Given that the location of a parent node *par* has already been chosen in the top-down processing, the location of its child node *v* is determined from two known quantities: (1) $|e_v|$, the edge length from *v* to its parent *par*, and (2) *ms(v)*, the locus of placements for *v* consistent with a minimum-cost ZST (lines 6-8). The location of *v*, i.e., *pl(v)*, can be determined as illustrated in Fig. 7.19.

Possible locations of *v*

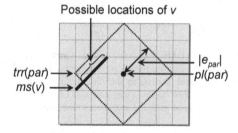

Fig. 7.19 Finding the location of child node *v* given the location of its parent node *par*.

Thus, the embeddings of all internal nodes of the topology can be determined, top-down, in linear time (Fig. 7.20).

The DME algorithm can also be applied to the BST (bounded-skew tree) problem [7.5], given a straightforward generalization from merging *segments* to merging *regions*. Each merging region in the bounded-skew DME construction is bounded by at most six segments having slopes +1, 0, -1, or +∞ (one of four possibilities). A related exercise is given at the end of this chapter.

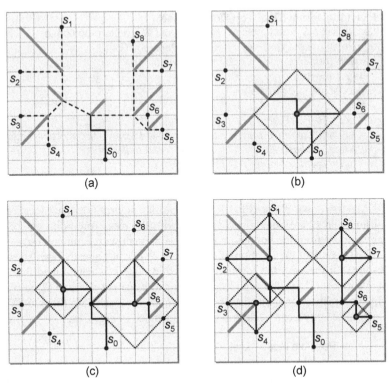

Fig. 7.20 Embedding the clock tree during the DME top-down phase. Gray lines indicate merging segments, dotted lines show connections between merging segments, and black lines indicate routing segments. (a) Connecting the clock source to the root merging segment. (b) Connecting the root merging segment to its children merging segments. (c) Connecting those merging segments to its children. (d) Connecting those merging segments to the sinks.

▶ **7.5.2 Clock Tree Buffering in the Presence of Variation**

High-performance designs require low clock skew, while their clock networks contribute significantly to power consumption. Therefore, tradeoffs between skew and total capacitance are important. Skew optimization requires highly accurate timing analysis. Simulation-based timing analysis is often too time-consuming, while closed-form delay models are too inaccurate. A commonly used compromise is to first perform optimization based on the Elmore delay model and then tune the tree with more accurate optimizations. For example, a difference of 5 picoseconds (ps) is only a 1% error relative to a 500 ps sink latency, but is a 50% error relative to a 10ps skew in the same tree. To address challenging skew constraints, a clock tree undergoes several optimization steps, including (1) *geometric clock tree construction* (Sec. 7.5.1), (2) initial *clock buffer insertion,* (3) *clock buffer sizing,* (4) *wire sizing,* and (5) *wire snaking.* In the presence of *process, voltage, and temperature* (*PVT*) variations, such optimizations require accurate models of the impacts of these variations.

High-level skew optimization. In the 1980s, the entire clock tree could be driven by a single buffer. However, with technology scaling, wires have become more resistive and clock trees today can no longer be driven by a single buffer. Therefore, clock buffers are inserted at multiple locations in the tree to ensure that the clock signal has sufficient strength to propagate to all sinks (timing points). The locations and sizes of buffers are used to control the propagation delay within each branch of the tree. Though not intended for clock trees, the algorithm proposed by *L. van Ginneken* [7.8] optimally buffers a tree to minimize Elmore delay between the source and every sink in $O(n^2)$ time, where n is the total number of possible buffer locations. The $O(n \log n)$-time variant proposed in [7.16] is more scalable. These algorithms also avoid insertion of unnecessary buffers on fast paths, thus achieving lower skew if the initial tree was balanced. After initial buffer insertion, subsequent optimizations are performed to minimize skew, decrease power consumption, and improve robustness of the clock tree to unforeseen changes to buffer characteristics, e.g., manufacturing process variations.

Clock buffer sizing. The choice of the clock buffer size for initial buffer insertion affects downstream optimizations, as most of the buffers' sizes and locations are unlikely to change. However, the best-performing size is difficult to identify analytically. Therefore, it is often determined by trial and error, e.g., using binary search. Clock buffer sizes can be further adjusted as follows. For a pair of sinks s_1 and s_2 with significant skew, find the unique path π in the tree connecting s_1 and s_2. Upsize the buffers on π according to a pre-computed table (discussed below) that matches an appropriate buffer size to each path length and fanout. In practice, larger buffers improve the robustness of the circuit, but consume more power and may introduce additional delay.

Wire sizing. The choice of wire width affects both power and susceptibility to manufacturing defects. Wider wires are more resilient to variation, but have greater capacitance and consume more dynamic power than thinner wires. Wider wires (and wider spacings to neighbors) are preferred for high-performance designs; thinner wires are preferred for low-power or less aggressive designs. After the initial wire width is chosen, it can be adjusted for individual segments based on timing analysis.

Low-level skew optimization. Compared to the global impact of high-level skew optimizations, low-level skew optimizations cause smaller, localized changes. The precision of low-level skew optimizations is typically much greater than that of high-level skew optimizations. Low-level optimizations, such as *wire sizing* and *wire snaking*, are preferred for fine-tuning skew. To slow down fast sinks, the length of the path can be increased by purposely detouring the wire. This increases the total capacitance and resistance of the path, thus increasing the propagation delay.

Variation modeling. Due to randomness in the semiconductor manufacturing process, every transistor in every chip is slightly different. In addition, every chip can be used at different ambient temperature, and will locally heat up or cool down depending on activity patterns. Supply voltage may also change depending on manufacturing variation and power drawn by other parts of the chip. Nevertheless, modern clock trees must operate as expected under a variety of circumstances. To ensure such robustness, an efficient and accurate *variation model* encapsulates the different parameters, e.g., wire width and thickness, of each library element as well-defined random variables. However, predicting the impact of process variations is difficult. One option is to run a large number of individual simulations with different parameter settings (*Monte-Carlo* simulation), but this is slow and impractical in an optimization flow.

A second option is to generate a lookup table that captures worst-case skew variations between pairs of sinks based on (1) technology node, (2) clock buffer and wire library, (3) tree path length, (4) variation model, and (5) desired yield. Though creating this table requires extensive simulations, this only needs to be done once for a given technology. The resulting table can be used for any compatible clock tree optimization, e.g., for clock buffer sizing, as previously explained in this section. In general, this lookup table approach facilitates a fast and accurate optimization.

Further clock network design techniques are discussed in Chaps. 42-43 of [7.1], including *active deskewing* and *clock meshes*, common in modern CPUs, as well as *clock gating*, used to decrease clock power dissipation. The book [7.18] focuses on clocking in modern VLSI systems from a designer perspective and recommends a number of techniques to minimize the impact of process variations. The book [7.15] discusses clocking for high-performance and low-power applications.

Chapter 7 Exercises

Exercise 1: Net Ordering

Recall from Chap. 5 that negotiated-congestion routing (NCR) allows for temporary routing violations and maintains history costs for every routing segment. These history costs can decrease the significance of net ordering. Given that NCR can be adapted to track-based detailed routing, give several reasons why net ordering remains important in area routing.

Exercise 2: Octilinear Maze Search

Compare octilinear maze search (Sec. 7.3.2) to breadth-first search (BFS) and Dijkstra's algorithm in terms of input data, output, and runtime. Can BFS or Dijkstra's algorithm be used instead of octilinear maze search?

Exercise 3: Method of Means and Medians (MMM)

Given the following locations of a clock source s_0 and eight sinks s_1-s_8:
(a) Draw the clock tree generated by MMM.
(b) Draw the clock tree topology generated by MMM.
(c) Calculate the total wirelength (in terms of grid units) and the clock skew using the linear delay model.

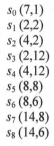

s_0 (7,1)
s_1 (2,2)
s_2 (4,2)
s_3 (2,12)
s_4 (4,12)
s_5 (8,8)
s_6 (8,6)
s_7 (14,8)
s_8 (14,6)

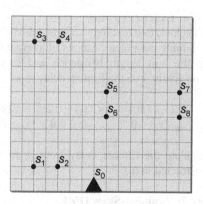

Exercise 4: Recursive Geometric Matching (RGM) Algorithm

Consider the same locations of a clock source s_0 and eight sinks s_1-s_8 as in Exercise 3.
(a) Draw the clock tree generated by RGM. Use the linear delay model when constructing tapping points. Let each sink have no delay, i.e., $t_{LD}(s_i) = 0$, where $1 \le i \le 8$.
(b) Draw the clock-tree topology generated by RGM.
(c) Calculate total wirelength (in terms of grid units) and clock skew using the linear delay model.

Exercise 5: Deferred-Merge Embedding (DME) Algorithm

Given the locations of a source s_0 and four sinks s_1-s_4 (left) and the following connection topology (middle), execute the bottom-up and top-down phases of DME, as described in Sec. 7.5.1.

(a) Top-down phase: draw the merging segments for the internal nodes, $ms(u_1)$, $ms(u_2)$ and $ms(u_3)$.

(b) Bottom-up phase: draw a minimum-cost embedded clock tree from s_0 to all four sinks s_1-s_4.

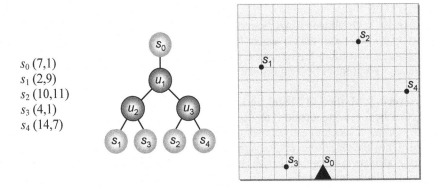

s_0 (7,1)
s_1 (2,9)
s_2 (10,11)
s_3 (4,1)
s_4 (14,7)

Exercise 6: Exact Zero Skew

Using the clock-tree topology and generated clock tree from Exercise 5:

(a) Determine the exact locations of internal nodes u_1, u_2 and u_3 to have zero skew. Let the unit length resistance and capacitance be $\alpha = 0.1$ and $\beta = 0.01$, respectively. Use the Elmore delay model with the following sink information.

$$C(s_1) = 0.1 \qquad C(s_2) = 0.2 \qquad C(s_3) = 0.3 \qquad C(s_4) = 0.2$$
$$t_{ED}(s_1) = 0 \qquad t_{ED}(s_2) = 0 \qquad t_{ED}(s_3) = 0 \qquad t_{ED}(s_4) = 0$$

(b) Calculate the Elmore delay from u_1 to each sink s_1-s_4 in the zero skew tree constructed in (a).

Exercise 7: Bounded-Skew DME

Given the locations of a source s_0 and four sinks s_1-s_4 (left), and using the following topology (center), calculate the bounded-skew feasible region for internal nodes, u_1, u_2 and u_3, using the linear delay model. The skew bound is four grid edges.

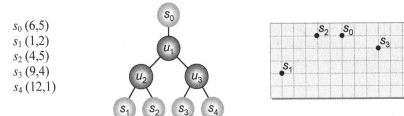

s_0 (6,5)
s_1 (1,2)
s_2 (4,5)
s_3 (9,4)
s_4 (12,1)

Chapter 7 References

[7.1]   C. J. Alpert, D. P. Mehta and S. S. Sapatnekar, eds., *Handbook of Algorithms for Physical Design Automation*, CRC Press, 2009.

[7.2]   H. B. Bakoglu, *Circuits, Interconnections and Packaging for VLSI*, Addison-Wesley, 1990.

[7.3]   K. D. Boese and A. B. Kahng, "Zero-Skew Clock Routing Trees with Minimum Wirelength", *Proc. Intl. Conf. on ASIC*, 1992, pp. 1.1.1-1.1.5.

[7.4]   T.-H. Chao, J.-M. Ho and Y.-C. Hsu, "Zero Skew Clock Net Routing", *Proc. Design Autom. Conf.*, 1992, pp. 518-523.

[7.5]   J. Cong, A. B. Kahng, C.-K. Koh and C.-W. A. Tsao, "Bounded-Skew Clock and Steiner Routing", *ACM Trans. on Design Autom. of Electronic Sys.* 3(3) (1998), pp. 341-388.

[7.6]   M. Edahiro, "An Efficient Zero-Skew Routing Algorithm", *Proc. Design Autom. Conf.*, 1994, pp. 375-380.

[7.7]   J. P. Fishburn, "Clock Skew Optimization", *IEEE Trans. on Computers* 39(7) (1990), pp. 945-951.

[7.8]   L. P. P. P. van Ginneken, "Buffer Placement in Distributed RC-Tree Networks for Minimal Elmore Delay", *Proc. Intl. Symp. on Circuits and Sys.*, 1990, pp. 865-868.

[7.9]   T.-Y. Ho, C.-F. Chang, Y.-W. Chang and S.-J. Chen, "Multilevel Full-Chip Routing for the X-Based Architecture", *Proc. Design Autom. Conf.*, 2005, pp. 597-602.

[7.10] M. A. B. Jackson, A. Srinivasan and E. S. Kuh, "Clock Routing for High-Performance ICs", *Proc. Design Autom. Conf.*, 1990, pp. 573-579.

[7.11] A. B. Kahng and G. Robins, *On Optimal Interconnections for VLSI*, Kluwer Academic Publishers, 1995.

[7.12] A. B. Kahng, J. Cong and G. Robins, "High-Performance Clock Routing Based on Recursive Geometric Matching", *Proc. Design Autom. Conf.*, 1991, pp. 322-327.

[7.13] C. Y. Lee, "An Algorithm for Path Connection and Its Applications", *IRE Trans. on Electronic Computers* 10 (1961), pp. 346-365.

[7.14] D. Lee and I. L. Markov, "CONTANGO: Integrated Optimization for SoC Clock Networks", *Proc. Design, Autom. and Test in Europe*, 2010, pp. 1468-1473.

[7.15] V. G. Oklobdzija, V. M. Stojanovic, D. M. Markovic and N. M. Nedovic, *Digital System Clocking: High-Performance and Low-Power Aspects*, Wiley-IEEE Press, 2003.

[7.16] W. Shi and Z. Li, "A Fast Algorithm for Optimal Buffer Insertion", *IEEE Trans. on CAD* 24(6) (2005), pp. 879-891.

[7.17] R.-S. Tsay, "Exact Zero Skew", *Proc. Intl. Conf. on CAD*, 1991, pp. 336-339.

[7.18] T. Xanthopoulos, ed., *Clocking in Modern VLSI Systems*, Springer, 2009.

[7.19] Q. Zhu, H. Zhou, T. Jing, X. Hong and Y. Yang, "Efficient Octilinear Steiner Tree Construction Based on Spanning Graphs", *Proc. Asia and South Pacific Design Autom. Conf.*, 2004, pp. 687-690.

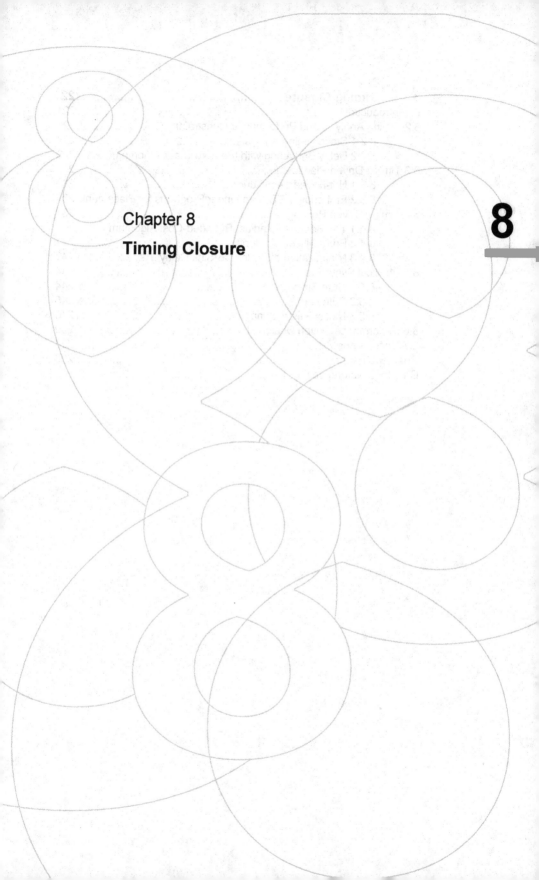

Chapter 8

Timing Closure

8

8 Timing Closure

The layout of an integrated circuit (IC) must not only satisfy geometric requirements, e.g., non-overlapping cells and routability, but also meet the design's *timing constraints*, e.g., *setup* (long-path) and *hold* (short-path) constraints. The optimization process that meets these requirements and constraints is often called *timing closure*. It integrates point optimizations discussed in previous chapters, such as placement (Chap. 4) and routing (Chaps. 5-7), with specialized methods to improve circuit performance. The following components of timing closure are covered in this chapter.

— *Timing-driven placement* (Sec. 8.3) minimizes signal delays when assigning locations to circuit elements.
— *Timing-driven routing* (Sec. 8.4) minimizes signal delays when selecting routing topologies and specific routes.
— *Physical synthesis* (Sec. 8.5) improves timing by changing the netlist.
  — Sizing transistors or gates: increasing the width:length ratio of transistors to decrease the delay or increase the drive strength of a gate (Sec. 8.5.1).
  — Inserting buffers into nets to decrease propagation delays (Sec. 8.5.2).
  — Restructuring the circuit along its critical paths (Sec. 8.5.3).

Sec. 8.6 integrates these optimizations in a *performance-driven physical design flow*.

8.1 Introduction

For many years, *signal propagation delay* in *logic gates* was the main contributor to circuit delay, while wire delay was negligible. Therefore, cell placement and wire routing did not noticeably affect circuit performance. Starting in the mid-1990s, technology scaling significantly increased the relative impact of wiring-induced delays, making high-quality placement and routing critical for timing closure.

Timing optimization engines must estimate circuit delays quickly and accurately to improve circuit timing. Timing optimizers adjust propagation delays through circuit components, with the primary goal of satisfying *timing constraints*, including

— *Setup* (*long-path*) *constraints*, which specify the amount of time a data input signal should be *stable* (steady) *before* the clock edge for each storage element (e.g., flip-flop or latch).
— *Hold-time* (*short-path*) *constraints*, which specify the amount of time a data input signal should be stable *after* the clock edge at each storage element.

Setup constraints ensure that no signal transition occurs too late. Initial phases of timing closure focus on these types of constraints, which are formulated as follows.

$$t_{cycle} \geq t_{combDelay} + t_{setup} + t_{skew}$$

Here, t_{cycle} is the clock period, $t_{combDelay}$ is the longest path delay through combinational logic, t_{setup} is the setup time of the receiving storage element (e.g., flip-flop), and t_{skew} is the *clock skew* (Sec. 7.4). Checking whether a circuit meets setup constraints requires estimating how long signal transitions will take to propagate from one storage element to the next. Such delay estimation is typically based on *static timing analysis* (*STA*), which propagates *actual arrival times* (*AATs*) and *required arrival times* (*RATs*) to the pins of every gate or cell. STA quickly identifies *timing violations*, and diagnoses them by tracing out *critical paths* in the circuit that are responsible for these timing failures (Sec. 8.2.1).

Motivated by efficiency considerations, STA does not consider circuit functionality and specific signal transitions. Instead, STA assumes that every cell propagates every 0-1 (1-0) transition from its input(s) to its output, and that every such propagation occurs with the worst possible delay[1]. Therefore, STA results are often pessimistic for large circuits. This pessimism is generally acceptable during optimization because it affects competing layouts equally, without biasing the optimization toward a particular layout. It is also possible to evaluate the timing of several competing layouts with more accurate techniques in order to choose the best solution.

One approach to mitigate pessimism in STA is to analyze the most critical paths. Some of these can be *false paths* – those that cannot be sensitized by any input transition because of the logic functions implemented by the gates or cells. IC designers often enumerate false paths that are likely to become timing-critical to exclude them from STA results and ignore them during timing optimization.

STA results are used to estimate how important each cell and each net are in a particular layout. A key metric for a given timing point g – that is, a pin of a gate or cell – is *timing slack*, the difference between g's RAT and AAT: $slack(g) = RAT(g) - AAT(g)$. Positive slack indicates that timing is met – the signal arrives before it is required – while negative slack indicates that timing is violated – the signal arrives after its required time. Algorithms for timing-driven layout guide the placement and routing processes according to timing slack values.

Guided by slack values, *physical synthesis* restructures the netlist to make it more suitable for high-performance layout implementation. For instance, given an unbalanced tree of gates, (1) the gates that lie on many critical paths can be upsized to propagate signals faster, (2) buffers may be inserted into long critical wires, and (3) the tree can be restructured to decrease its overall depth.

---

[1] Path-based approaches for timing optimizations are discussed in Secs. 8.3-8.4.

Hold-time constraints ensure that signal transitions do not occur too early. Hold violations can occur when a signal path is too short, allowing a receiving flip-flop to capture the signal at the current cycle instead of the next cycle. The hold-time constraint is formulated as follows.

$$t_{combDelay} \geq t_{hold} + t_{skew}$$

Here, $t_{combDelay}$ is the delay of the circuit's combinational logic, t_{hold} is the hold time required for the receiving storage element, and t_{skew} is the clock skew. As clock skew affects hold-time constraints significantly more than setup constraints, hold-time constraints are typically enforced after synthesizing the clock network (Sec. 7.4).

Timing closure is the process of satisfying timing constraints through layout optimizations and netlist modifications. It is common to use verbal expressions such as "the design has closed timing" when the design satisfies all timing constraints.

This chapter focuses on individual timing algorithms (Secs. 8.2-8.4) and optimizations (Sec. 8.5), but in practice, these must be applied in a carefully balanced design flow (Sec. 8.6). Timing closure may repeatedly invoke certain optimization and analysis steps in a loop until no further improvement is observed. In some cases, the choice of optimization steps depends on the success of previous steps as well as the distribution of timing slacks computed by STA.

8.2 Timing Analysis and Performance Constraints   8.2

Almost all digital ICs are *synchronous, finite state machines (FSM)*, or *sequential machines*. In FSMs, transitions occur at a set clock frequency. Fig. 8.1 "unrolls" a sequential circuit in time, from one clock period to the next. The figure shows two types of circuit components: (1) clocked *storage elements*, e.g., flip-flops or latches, also referred to as *sequential elements*, and (2) *combinational logic*. During each clock period in the operation of the sequential machine, (1) present-state bits stored in the clocked storage elements flow from the storage elements' output pins, along with system inputs, into the combinational logic, (2) the network of combinational logic then generates values of next-state functions, along with system outputs, and (3) the next-state bits flow into the clocked elements' data input pins, and are stored at the next clock tick.

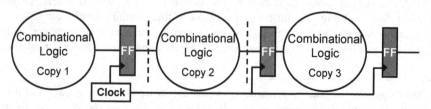

Fig. 8.1 A sequential circuit, consisting of flip-flops and combinational logic, "unrolled" in time.

The maximum clock frequency for a given design depends upon (1) *gate delays*, which are the signal delays due to gate transitions, (2) *wire delays*, which are the delays associated with signal propagation along wires, and (3) *clock skew* (Sec. 7.4). In practice, the predominant sources of delay in standard signals come from gate and wire delays. Therefore, when analyzing setup constraints, this section considers clock skew negligible. A lower bound on the design's clock period is given by the sum of gate and wire delays along any *timing path* through combinational logic – from the output of a storage element to the input of the next storage element. This lower bound on the clock period determines an upper bound on the clock frequency.

In earlier technologies, gate delays accounted for the majority of circuit delay, and the number of gates on a timing path provided a reasonable estimate of path delay. However, in recent technologies, wire delay, along with the component of gate delay that is dependent on capacitive loading, comprises a substantial portion of overall path delay. This adds complexity to the task of estimating path delays and, hence, achievable (maximum) clock frequency.

For a chip to function correctly, *path delay constraints* (Sec. 8.3.2) must be satisfied whenever a signal transition traverses a path through combinational logic. The most critical verification task faced by the designer is to confirm that all path delay constraints are satisfied. To do this *dynamically*, i.e., using circuit simulation is infeasible for two reasons. First, it is computationally intractable to enumerate all possible combinations of state and input variables that can cause a transition, i.e., *sensitize*, a given combinational logic path. Second, there can be an exponential number of paths through the combinational logic. Consequently, design teams often *signoff* on circuit timing *statically*, using a methodology that pessimistically assumes all combinational logic paths can be sensitized. This framework for timing closure is based on *static timing analysis (STA)* (Sec. 8.2.1), an efficient, linear-time verification process that identifies *critical paths*.

After critical paths have been identified, *delay budgeting*[2] (Secs. 8.2.2 and 8.3.1) sets upper bounds on the lengths or propagation delays for these paths, e.g., using the *zero-slack algorithm* [8.19], which is covered in Sec. 8.2.2. Other delay budgeting techniques are described in [8.29].

▶ **8.2.1 Static Timing Analysis**

In STA, a combinational logic network is represented as a *directed acyclic graph (DAG)* (Sec. 1.7). Fig. 8.2 illustrates a network of four combinational logic gates x, y, z and w, three inputs a, b and c, and one output f. The inputs are annotated with times 0, 0 and 0.6 time units, respectively, at which signal transitions occur relative

---

[2] This methodology is intended for layout of circuits directly represented by graphs rather than circuits partitioned into high-level modules. However, this methodology can also be adapted to assign budgets to entire modules instead of circuit elements.

to the start of the clock cycle. Fig. 8.2 also shows gate and wire delays, e.g., the gate delay from the input to the output of inverter x is 1 unit, and the wire delay from input b to the input of inverter x is 0.1 units. For modern designs, gate and wire delays are typically on the order of *picoseconds* (*ps*).

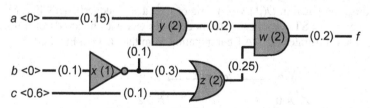

Fig. 8.2 Three inputs a, b and c are annotated with times at which signal transitions occur in angular brackets. Each edge and gate is annotated with its delay in parentheses.

Fig. 8.3 illustrates the corresponding DAG, which has one node for each input and output, as well as one node for each logic gate. For convenience, a source node is introduced with a directed edge to each input. Nodes corresponding to logic gates are labeled with the respective gate delays (e.g., node y has the label 2). Directed edges from the source to the inputs are labeled with transition times, and directed edges between gate nodes are labeled with wire delays. Because this DAG representation has one node per logic gate, it follows the *gate node convention*. The *pin node convention*, where the DAG has a node for each pin of each gate, is more detailed and will be used later in this section.

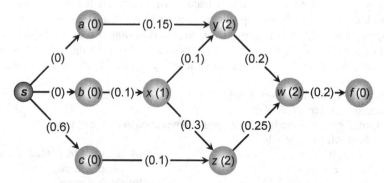

Fig. 8.3 DAG representation using gate node convention of the circuit in Fig. 8.2.

Actual arrival time. In a circuit, the latest transition time at a given node $v \in V$, measured from the beginning of the clock cycle, is the *actual arrival time* (*AAT*), denoted as $AAT(v)$. By convention, this is the arrival time at the *output side* of node v. For example, in Fig. 8.3, $AAT(x) = 1.1$ because of the wire delay from input b (0.1) and the gate delay of inverter x (1.0). For node y, although the signal transitions on the path through a will arrive at time 2.15, the arrival time is dominated by transitions along the path through x. Hence, $AAT(a) = 3.2$. Formally, the AAT of node v is

$$AAT(v) = \max_{u \in FI(v)} \big(AAT(u) + t(u,v) \big)$$

where $FI(v)$ is the set of all nodes from which there exists a directed edge to v, and $t(u,v)$ is the delay on the (u,v) edge. This recurrence enables all AAT values in the DAG to be computed in $O(|V| + |E|)$ time or $O(|gates| + |edges|)$. This linear scaling of runtime makes STA applicable to modern designs with hundreds of millions of gates. Fig 8.4 illustrates the AAT computation from the DAG in Fig. 8.3.

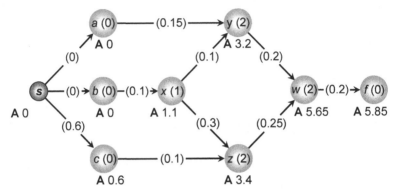

Fig. 8.4 Actual arrival times (AATs) of the DAG, denoted with 'A', of Fig. 8.3.

Although STA is *pessimistic*, i.e., the longest path in the DAG of a circuit might not actually be sensitizable, the design team must satisfy all timing constraints. An algorithm to find all longest paths from the source in a DAG was proposed by *Kirkpatrick* in 1966 [8.18]. It uses a *topological ordering* of nodes – if there exists a (u,v) edge in the DAG, then u is ordered before v. This ordering can be determined in linear time by reversing the post-order labeling obtained by depth-first search.

Longest Paths Algorithm [8.18]
Input: directed graph $G(V,E)$
Output: AATs of all nodes $v \in V$ based on worst-case (longest) paths
1. **foreach** (node $v \in V$)
2. $AAT[v] = -\infty$ // all AATs are by default unknown
3.   AAT[*source*] = 0   //   except *source*, which is 0
4.   Q = TOPO_SORT(V)   // topological order
5. **while** ($Q \neq \varnothing$)
6. u = FIRST_ELEMENT(Q)   // u is the first element in Q
7.   **foreach** (neighboring node v of u)
8. $AAT[v]$ = MAX($AAT[v], AAT[u] + t[u][v]$) // $t[u][v]$ is the (u,v) edge delay
9.   REMOVE(Q,u)   // remove u from Q

Required arrival time. The *required arrival time* (*RAT*), denoted as $RAT(v)$, is the time by which the latest transition at a given node v must occur in order for the circuit to operate correctly within a given clock cycle. Unlike AATs, which are determined from multiple paths from *upstream* inputs and flip-flop outputs, RATs are determined from multiple paths to *downstream* outputs and flip-flop inputs. For example, suppose that $RAT(f)$ for the circuit in Fig 8.2 is 5.5. This forces $RAT(w)$ to be 5.3, $RAT(y)$ to be 3.1, and so on (Fig. 8.5). Formally, the RAT of a node v is

$$RAT(v) = \max_{u \in FO(v)} \left(RAT(u) - t(u,v)\right)$$

where $FO(v)$ is the set of all nodes with a directed edge from v.

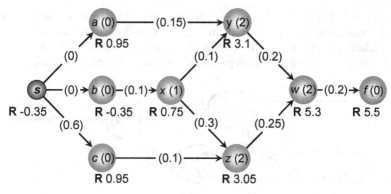

Fig. 8.5 Required arrival times (RATs) of the DAG, denoted with 'R', of Fig. 8.3.

Slack. Correct operation of the chip with respect to *setup constraints*, e.g., maximum path delay, requires that the AAT at each node does not exceed the RAT. That is, for all nodes $v \in V$, $AAT(v) \leq RAT(v)$ must hold.

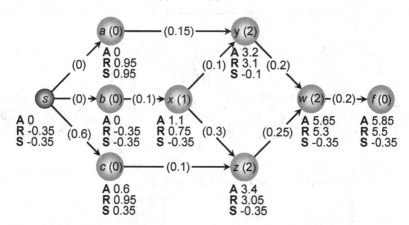

Fig. 8.6 The STA result of the DAG of Fig. 8.3, showing the actual (**A**) and required (**R**) arrival times, and the slack (**S**) of each node.

The *slack* of a node v, defined as

$$slack(v) = RAT(v) - AAT(v)$$

is an indicator of whether the timing constraint for v has been satisfied. *Critical paths* or *critical nets* are signals that have *negative slack*, while *non-critical paths* or *non-critical nets* have *positive slack*.

Timing optimization (Sec. 8.5) is the process by which (1) negative slack is increased to achieve design correctness, and (2) positive slack is reduced to minimize overdesign and recover power and area. Fig 8.6 illustrates the full STA computation, including slack, from Fig. 8.3.

A DAG labeled with the *pin node convention* facilitates a more detailed and accurate timing analysis, as the delay of a gate output depends on which input pin has switched. Fig. 8.7 shows the circuit of Fig. 8.2 annotated with the pin node convention, where v_i is the ith input pin of gate v, and v_O is the output pin of v.

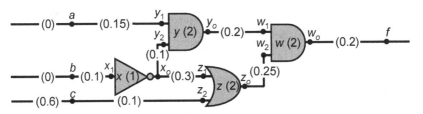

Fig. 8.7 Circuit of Fig. 8.2 annotated with the pin node convention. For a logic gate v, v_i denotes its ith input pin, and v_O denotes its output pin.

Fig. 8.8 shows the result of STA constructed using the pin node convention.

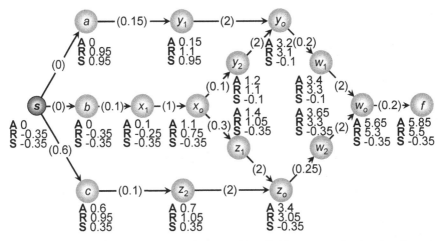

Fig. 8.8 STA result for the circuit of Fig. 8.2, with a DAG constructed using the pin node convention. Each node in the DAG is an input pin or an output pin of a logic gate.

Current practice. In modern designs, separate timing analyses are performed for the cases of *rise delay* (rising transitions) and *fall delay* (falling transitions).

Signal integrity extensions to STA consider changes in delay due to switching activity on neighboring wires of the path under analysis. For signal integrity analysis, the STA engine keeps track of *windows* (*intervals*) of AATs and RATs, and typically executes multiple timing analysis iterations before these timing windows stabilize to a clean and accurate result.

Statistical STA (*SSTA*) is a generalization of STA where gate and wire delays are modeled by random variables and represented by probability distributions [8.21]. Propagated AATs, RATs and timing slacks are also random variables. In this context, timing constraints can be satisfied with high probability (e.g., 95%). SSTA is an increasingly popular methodology choice for leading-edge designs, due to the increased manufacturing variability in advanced process nodes. Propagating statistical distributions instead of intervals avoids some of STA's inherent pessimism. This reduces the costly power, area and schedule impacts of overdesign.

The static verification approach is continually challenged by two fundamental weaknesses – (1) the assumption of a clock and (2) the assumption that all paths are sensitizable. First, STA is not applicable in *asynchronous* contexts, which are increasingly prevalent in modern designs, e.g., asynchronous interfaces in systems-on-chips (SoCs), asynchronous logic design styles to improve speed and power. Second, optimization tools waste considerable runtime and chip resources – e.g., power, area and speed – satisfying "phantom" constraints. In practice, designers can manually or semi-automatically specify *false* and *multicycle paths* – paths whose signal transitions do not need to finish within one clock cycle. Methodologies to fully exploit the availability of such *timing exceptions* are still under development.

► 8.2.2 Delay Budgeting with the Zero-Slack Algorithm

In timing-driven physical design, both gate and wire delays must be optimized to obtain a timing-correct layout. However, there is a chicken-and-egg dilemma: (1) timing optimization requires knowledge of capacitive loads and, hence, actual wirelength, but (2) wirelengths are unknown until placement and routing are completed. To help resolve this dilemma, *timing budgets* are used to establish delay and wirelength constraints for each net, thereby guiding placement and routing to a timing-correct result. The best-known approach to timing budgeting is the *zero-slack algorithm* (*ZSA*) [8.19], which is widely used in practice.

Algorithm. Consider a netlist consisting of logic gates $v_1, v_2, \dots, v_n$ and nets $e_1, e_2, \dots, e_n$, where e_i is the output net of gate v_i. Let $t(v)$ be the gate delay of v, and let $t(e)$ be the wire delay of e.[3] The ZSA takes the netlist as input, and seeks to decrease positive slacks of all nodes to zero by increasing $t(v)$ and $t(e)$ values. These increased delay values together constitute the *timing budget* $TB(v)$ of node v, which should not be exceeded during placement and routing.

---

[3] A multi-fanout net e_i has multiple source-sink delays, so ZSA must be adjusted accordingly.

$$TB(v) = t(v) + t(e)$$

If $TB(v)$ is exceeded, then the place-and-route tool typically (1) decreases the wirelength of e by replacement or rerouting, or (2) changes the size of gate v. The delay impact of a wire or gate size change can be estimated using the *Elmore delay* model [8.13]. If most *arcs* (branches) of a timing path are within budget, then the path may meet its timing constraints even if some arcs exceed their budgets. Thus, another approach to satisfying the timing budget is (3) *rebudgeting*. As in Sec. 8.2.1, let $AAT(v)$, $RAT(v)$ and $slack(v)$ denote respectively the AAT, RAT, and slack at node v in the timing graph G.

Zero-Slack Algorithm (Late-Mode Analysis)
Input: timing graph $G(V,E)$
Output: timing budgets TB for each $v \in V$
1.   **do**
2. $(AAT,RAT,slack)$ = STA(G)
3.   **foreach** $(v_i \in V)$
4. $TB[v_i]$ = DELAY(v_i) + DELAY(e_i)
5. $slack_{min}$ = ∞
6.   **foreach** $(v \in V)$
7.   **if** $((slack[v] < slack_{min})$ **and** $(slack[v] > 0))$
8. $slack_{min}$ = $slack[v]$
9. v_{min} = v
10.   **if** $(slack_{min} \neq \infty)$
11. $path$ = v_{min}
12.   ADD_TO_FRONT($path$,BACKWARD_PATH(v_{min},G))
13.   ADD_TO_BACK($path$,FORWARD_PATH(v_{min},G))
14. s = $slack_{min}$ / $|path|$
15.   **for** $(i = 1$ to $|path|)$
16. $node$ = $path[i]$ // evenly distribute
17. $TB[node]$ = $TB[node]$ + s // slack along $path$
18.   **while** $(slack_{min} \neq \infty)$

The ZSA consists of three major steps. First, determine the initial slacks of all nodes (lines 2-4), and select a node v_{min} with minimum positive slack $slack_{min}$ (lines 5-9). Second, find a path $path$ of nodes that *dominates* $slack(v_{min})$, i.e., any change in delays in $path$'s nodes will cause $slack(v_{min})$ to change. This is done by calling the two procedures *BACKWARD_PATH* and *FORWARD_PATH* (lines 12-13). Third, evenly distribute the slack by increasing $TB(v)$ for each v in $path$ (lines 14-17). Each budget increment s will decrement a node slack $slack(v)$. By repeating the process (lines 1-18), the slack of each node in V will end up at zero. The resulting timing budgets at all nodes are the final output of ZSA. Further details of ZSA, including proofs of correctness and complexity analyses, are given in [8.19].

FORWARD_PATH(v_{min},G) constructs a path starting with node v_{min}, and iteratively adds a node v to the path from among the fanouts of the previously-added node in $path$ (lines 2-10). Each node v satisfies the condition in line 6, where $RAT(v_{min})$ is determined by $RAT(v)$, and $AAT(v)$ is determined by $AAT(v_{min})$ – changing the delay of either node affects the slack of both nodes.

Forward Path Search (FORWARD_PATH(v_{min},G))
Input: node v_{min} with minimum slack $slack_{min}$, timing graph G
Output: maximal downstream path $path$ from v_{min} such that no node $v \in V$ affects
        the slack of $path$
1. $path = v_{min}$
2.   **do**
3. $flag$ = **false**
4. $node$ = LAST_ELEMENT($path$)
5.   **foreach** (fanout node fo of $node$)
6.   **if** (($RAT[fo]$ == $RAT[node]$ + $TB[fo]$) **and** ($AAT[fo]$ == $AAT[node]$ + $TB[fo]$))
7.   ADD_TO_BACK($path$,fo)
8. $flag$ = **true**
9.   **break**
10. **while** ($flag$ == **true**)
11. REMOVE_FIRST_ELEMENT($path$)   // remove v_{min}

$BACKWARD\_PATH(v_{min},G)$ iteratively finds a fanin node of the current node so that both nodes' slacks will change if either node's delay is changed.

Backward Path Search (BACKWARD_PATH(v_{min},G))
Input: node v_{min} with minimum slack $slack_{min}$, timing graph G
Output: maximal upstream path $path$ from v_{min} such that no node $v \in V$ affects the
        slack of $path$
1. $path = v_{min}$
2.   **do**
3. $flag$ = **false**
4. $node$ = FIRST_ELEMENT($path$)
5.   **foreach** (fanin node fi of $node$)
6.   **if** (($RAT[fi]$ == $RAT[node]$ – $TB[fi]$) **and** ($AAT[fi]$ == $AAT[node]$ – $TB[fi]$))
7.   ADD_TO_FRONT($path$,fi)
8. $flag$ = **true**
9.   **break**
10. **while** ($flag$ == **true**)
11. REMOVE_LAST_ELEMENT($path$)   // remove v_{min}

Early-mode analysis. ZSA uses *late-mode analysis* with respect to *setup constraints*, i.e., the latest times by which signal transitions can occur for the circuit to operate correctly. Correct operation also depends on satisfying *hold-time constraints* on the earliest signal transition times. *Early-mode analysis* considers these constraints. If the data input of a sequential element changes too soon after the triggering edge of the clock signal, the logic value at the output of that sequential element may become incorrect during the current clock cycle. As geometries shrink, early-mode violations have become an overriding concern. While setup violations can be avoided by lowering the chip's operating frequency, the chip's cycle time does not affect hold-time constraints (Sec. 8.1). Violations of hold-time constraints typically result in chip failures.

To correctly analyze this timing constraint, the *earliest actual arrival time* of signal transitions at each node must be determined. The *required arrival time* of a sequential element *in early mode* is the time at which the earliest signal can arrive and still satisfy the library-cell hold-time requirement.

For each gate v, $AAT_{EM}(v) \geq RAT_{EM}(v)$ must be satisfied, where $AAT_{EM}(v)$ is the earliest actual arrival time of a signal transition, and $RAT_{EM}(v)$ is the required arrival time in early mode, at gate v. The early-mode slack is then defined as

$$slack_{EM}(v) = AAT_{EM}(v) - RAT_{EM}(v)$$

When adapted to early-mode analysis, ZSA is called the *near zero-slack algorithm*. The adapted algorithm seeks to *decrease* $TB(v)$ by decreasing $t(v)$ or $t(e)$, so that all nodes have minimum early-mode timing slacks. However, since $t(v)$ and $t(e)$ cannot be negative, node slacks may not necessarily all become zero.

Near Zero-Slack Algorithm (Early-Mode Analysis)
Input: timing graph $G(V,E)$
Output: timing budgets TB for each $v \in V$
1.   **foreach** (node $v \in V$)
2. $done[v]$ = **false**
3.   **do**
4. $(RAT_{EM}, AAT_{EM}, slack_{EM})$ = STA_EM(G)   // early-mode STA
5. $slack_{min}$ = ∞
6.   **foreach** (node $v_i \in V$)
7. $TB[v_i]$ = DELAY(v_i) + DELAY(e_i)
8.   **foreach** (node $v \in V$)
9.   **if** (($done[v]$ == **false**) **and** ($slack_{EM}[v]$ < $slack_{min}$) **and** ($slack_{EM}[v]$ > 0))
10. $slack_{min}$ = $slack_{EM}[v]$
11. v_{min} = v
12.   **if** ($slack_{min} \neq \infty$)
13. $path$ = v_{min}
14.   ADD_TO_FRONT($path$, BACKWARD_PATH_ EM(v_{min}, G))
15.   ADD_TO_BACK($path$, FORWARD_PATH _EM(v_{min}, G))
16.   **for** ($i = 1$ **to** $|path|$)
17. $node$ = $path[i]$
18. $path\_E[i]$ = FIND_EDGE($V[node], E$)   // corresponding edge of v_i
19.   **for** ($i = 1$ **to** $|path|$)
20. $node$ = $path[i]$
21. s = MIN($slack_{min}$ / $|path|$, DELAY($path\_E[i]$))
22. $TB[node]$ = $TB[node] - s$ // decrease DELAY($node$) or
                                                                // DELAY($path\_E[node]$)
23.   **if** (DELAY($path\_E[i]$) == 0)
24. $done[node]$ = **true**
25.   **while** ($slack_{min} < \infty$)

In relation to the ZSA pseudocode, the procedure $BACKWARD\_PATH\_EM(v_{min},G)$ is equivalent to $BACKWARD\_PATH(v_{min},G)$, and $FORWARD\_PATH\_EM(v_{min},G)$ is the equivalent to $FORWARD\_PATH(v_{min},G)$, except that early-mode analysis is used for all arrival and required times.

Compared to the original ZSA, the two differences are (1) the use of early-mode timing constraints and (2) the handling of the Boolean flag $done(v)$. Lines 1-2 set $done(v)$ to false for every node v in V. If $t(e_i)$ reaches zero for a node v_i, $done(v_i)$ is set to true (lines 23-24). In subsequent iterations (lines 3-25), if v_i is the minimum slack node of G, it will be skipped (line 9) because $t(e_i)$ cannot be decreased further. After the algorithm completes, each node v will either have $slack(v) = 0$ or $done(v) = $ true.

In practice, if the delay of a node does not satisfy its early-mode timing budget, the delay constraint can be satisfied by adding additional delay (padding) to appropriate components. However, there is always the danger that additional delay may cause violations of late-mode timing constraints. Thus, a circuit should be first designed with ZSA and late-mode analysis. Early-mode analysis may then be used to confirm that early-mode constraints are satisfied, or to guide circuit modifications to satisfy such constraints.

8.3 Timing-Driven Placement   8.3

Timing-driven placement (*TDP*) optimizes *circuit delay*, either to satisfy all *timing constraints* or to achieve the greatest possible clock frequency. It uses the results of STA (Sec. 8.2.1) to identify *critical nets* and attempts to improve signal propagation delay through those nets. Typically, TDP minimizes one or both of the following. (1) *worst negative slack* (*WNS*)

$$WNS = \min_{\tau \in T}\left(slack(\tau)\right)$$

where T is the set of timing endpoints, e.g., primary outputs and inputs to flip-flops, and (2) *total negative slack* (*TNS*)

$$TNS = \sum_{\tau \in T, slack(\tau)<0} slack(\tau)$$

Algorithmic techniques for timing-driven placement can be categorized as *net-based* (Sec. 8.3.1), *path-based* or *integrated* (Sec. 8.3.2). There are two types of net-based techniques – (1) *delay budgeting* assigns upper bounds to the timing or length of individual nets, and (2) *net weighting* assigns higher priorities to *critical nets* during placement. Path-based placement seeks to shorten or speed up entire *timing-critical* paths rather than individual nets. While more accurate than net-based placement, path-based placement does not scale to large, modern designs because the number of paths in some circuits, such as multipliers, can grow exponentially with the number of gates. Both path-based and net-based approaches (1) rely on support within the placement algorithm, and (2) require a dedicated infrastructure for (incremental) calculation of timing statistics and parameters. Some placement approaches facilitate integration with timing-driven techniques. For instance, net weighting is naturally supported by simulated annealing and all analytic algorithms. Netlist partitioning algorithms support small integer net weights, but can usually be extended to support non-integer weights, either by scaling or by replacing bucket-based data structures with more general priority queues.

Timing-driven placement algorithms often operate in multiple iterations, during which the delay budgets or net weights are adjusted based on the results of STA. Integrated algorithms typically use constraint-driven mathematical formulations in which STA results are incorporated as constraints and possibly in the objective function. Several TDP methods are discussed below, while more advanced algorithms can be found in [8.8], [8.17], [8.20], and Chap. 21 of [8.5].

In practice, some industrial flows do not incorporate timing-driven methods during initial placement because timing information can be very inaccurate until locations are available. Instead, subsequent placement iterations, especially during detailed placement, perform timing optimizations. Integrated methods are commonly used; for example, the linear programming formulation (Sec. 8.3.2) is generally more accurate than net-weighting or delay budgeting, at the cost of increased runtime. A practical design flow for timing closure is introduced in Sec. 8.6.

▶ **8.3.1 Net-Based Techniques**

Net-based approaches impose either quantitative priorities that reflect timing criticality (net weights), or upper bounds on the timing of nets, in the form of *net constraints* (delay budgets). Net weights are more effective at the early design stages, while delay budgets are more meaningful if timing analysis is more accurate. More information on net weighting can be found in [8.12].

Net weighting. Recall that a traditional placer optimizes total wirelength and routability. To account for timing, a placer can minimize the total *weighted* wirelength, where each net is assigned a net weight (Chap. 4). Typically, the higher the net weight is, the more timing-critical the net is considered. In practice, net weights are assigned either *statically* or *dynamically* to improve timing.

Static net weights are computed before placement and do not change. They are usually based on slack – the more critical the net (the smaller the slack), the greater the weight. Static net weights can be either *discrete*, e.g.,

$$w = \begin{cases} \omega_1 & \text{if } slack > 0 \\ \omega_2 & \text{if } slack \leq 0 \end{cases}, \text{ where } \omega_1 > 0, \omega_2 > 0, \text{ and } \omega_2 > \omega_1$$

where $\omega_1 < \omega_2$ are constants greater than zero, or *continuous*, e.g.,

$$w = \left(1 - \frac{slack}{t}\right)^{\alpha}$$

where t is the longest path delay and α is a criticality exponent.

In addition to slack, various other parameters can be accounted for, such as net size and the number of critical paths traversing a given net. However, assigning too many higher weights may lead to increased total wirelength, routability difficulties, and the emergence of new critical paths. In other words, excessive net weighting may eventually lead to inferior timing. To this end, net weights can be assigned based on *sensitivity*, or how each net affects TNS. For example, the authors of [8.27] define the net weight of *net* as follows. Let

- $w_o(net)$ be the original net weight of *net*
- $slack(net)$ be the slack of *net*
- $slack_{target}$ be the target slack of the design
- $s_w^{SLACK}(net)$ be the slack sensitivity to the net weight of *net*
- $s_w^{TNS}(net)$ be the TNS sensitivity to the net weight of *net*
- α and β be constant bounds on the net weight change that control the tradeoff between WNS and TNS

Then, if $slack(net) \leq 0$,

$$w(net) = w_o(net) + \alpha \cdot (slack_{target} - slack(net)) \cdot s_w^{SLACK}(net) + \beta \cdot s_w^{TNS}(net)$$

Otherwise, if $slack(net) > 0$, then $w(net)$ remains the same, i.e., $w(net) = w_o(net)$.

Dynamic net weights are computed during placement iterations and keep an updated timing profile. This can be more effective than static net weights, since they are computed before placement, and can become outdated when net lengths change. An example method updates slack values based on efficient calculation of incremental slack for each net *net* [8.7]. For a given iteration k, let

- $slack_{k-1}(net)$ be the slack at iteration $k - 1$
- $s_L^{DELAY}(net)$ be the delay sensitivity to the wirelength of *net*
- $\Delta L(net)$ be the change in wirelength between iteration $k - 1$ and k for *net*

Then, the estimated slack of *net* at iteration k is

$$slack_k(net) = slack_{k-1}(net) - s_L^{DELAY}(net) \cdot \Delta L(net)$$

After the timing information has been updated, the net weights should be adjusted accordingly. In general, this incremental method of weight modification is based on previous iterations. For instance, for each net *net*, the authors of [8.14] first compute the *net criticality* υ at iteration k as

$$\upsilon_k(net) = \begin{cases} \dfrac{1}{2}(\upsilon_{k-1}(net) + 1) & \text{if } net \text{ is among the 3\% most critical nets} \\ \dfrac{1}{2}\upsilon_{k-1}(net) & \text{otherwise} \end{cases}$$

and then update the net weights as

$$w_k(net) = w_{k-1}(net) \cdot (1 + \upsilon_k(net))$$

Variants include using the previous j iterations and using different relations between the net weight and criticality.

In practice, dynamic methods can be more effective than using static net weights, but require careful net weight assignment. Unlike static net weights, which are relevant to any placer, dynamic net weights are typically tailored to each type of placer; their computation is integrated with the placement algorithm. To be scalable, the re-computation of timing information and net weights must be efficient [8.7].

Delay budgeting. An alternative to using net weights is to limit the delay, or the total length, of each net by using *net constraints*. This mitigates several drawbacks of net weighting. First, predicting the exact effect of a net weight on timing or total wirelength is difficult. For example, increasing weights of multiple nets may lead to the same (or very similar) placement. Second, there is no guarantee that a net's timing or length will decrease because of a higher net weight. Instead, net-constraint methods have better control and explicitly limit the length or slack of nets. However, to ensure scalability, net constraints must be generated such that they do not over-constrain the solution space or limit the total number of solutions, thereby hurting solution quality. In practice, these net constraints can be generated *statically*, before placement, or *dynamically*, when the net constraints are added or modified during each iteration of placement. A common method to calculate delay budgets is the *zero-slack algorithm* (ZSA), previously discussed in Sec. 8.2.2. Other advanced methods for delay budgeting can be found in [8.15].

The support for constraints in each type of placer must be implemented carefully so as to not sacrifice runtime or solution quality. For instance, min-cut placers must choose how to assign cells to partitions while meeting wirelength constraints. To meet these constraints, some cells may have to be assigned to certain partitions. Force-directed placers can adjust the attraction force on certain nets that exceed a certain length, but must ensure that these forces are in balance with those forces on other nets. More advanced algorithms for min-cut and force-directed placers on TDP can be found in [8.16] and [8.26], respectively.

▶ 8.3.2 Embedding STA into Linear Programs for Placement

Unlike net-based methods, where the timing requirements are mapped to net weights or net constraints, path-based methods for timing-driven placement *directly* optimize the design's timing. However, as the number of (critical) paths of concern can grow quickly, this method is much slower than net-based approaches. To improve scalability, timing analysis may be captured by a *set of constraints* and an *optimization objective* within a mathematical programming framework, such as *linear programming*. In the context of timing-driven placement, a linear program (LP) minimizes a function of slack, such as TNS, subject to two major types of constraints: (1) *physical*, which define the locations of the cells, and (2) *timing*, which define the slack requirements. Other constraints such as electrical constraints may also be incorporated.

Physical constraints. The physical constraints can be defined as follows. Given the set of cells V and the set of nets E, let

- x_v and y_v be the center of cell $v \in V$
- V_e be the set of cells connected to net $e \in E$
- $left(e)$, $right(e)$, $bottom(e)$, and $top(e)$ respectively be the coordinates of the left, right, bottom, and top boundaries of e's bounding box
- $\delta_x(v,e)$ and $\delta_y(v,e)$ be pin offsets from x_v and y_v for v's pin connected to e

Then, for all $v \in V_e$,

$$left(e) \leq x_v + \delta_x(v,e)$$
$$right(e) \geq x_v + \delta_x(v,e)$$
$$bottom(e) \leq y_v + \delta_y(v,e)$$
$$top(e) \geq y_v + \delta_y(v,e)$$

That is, every pin of a given net e must be contained within e's bounding box. Then, e's *half-perimeter wirelength* (*HPWL*) (Sec. 4.2) is defined as

$$L(e) = right(e) - left(e) + top(e) - bottom(e)$$

Timing constraints. The timing constraints can be defined as follows. Let

- $t_{GATE}(v_i,v_o)$ be the gate delay from an input pin v_i to the output pin v_o for cell v
- $t_{NET}(e,u_o,v_i)$ be net e's delay from cell u's output pin u_o to cell v's input pin v_i
- $AAT(v_j)$ be the arrival time on pin j of cell v

Then, define two types of timing constraints – those that account for input pins, and those that account for output pins.

For every input pin v_i of cell v, the arrival time at each v_i is the arrival time at the previous output pin u_o of cell u plus the net delay.

$$AAT(v_i) = AAT(u_o) + t_{NET}(u_o, v_i)$$

For every output pin v_o of cell v, the arrival time at v_o should be greater than or equal to the arrival time plus gate delay of each input v_i. That is, for each input v_i of cell v,

$$AAT(v_o) \geq AAT(v_i) + t_{GATE}(v_i, v_o)$$

For every pin τ_p in a *sequential* cell τ, the slack is computed as the difference between the required arrival time $RAT(\tau_p)$ and actual arrival time $AAT(\tau_p)$.

$$slack(\tau_p) \leq RAT(\tau_p) - AAT(\tau_p)$$

The required time $RAT(\tau_p)$ is specified at every input pin of a flip-flop and all primary outputs, and the arrival time $AAT(\tau_p)$ is specified at each output pin of a flip-flop and all primary inputs. To ensure that the program does not over-optimize, i.e., does not optimize beyond what is required to (safely) meet timing, upper bound all pin slacks by zero (or a small positive value).

$$slack(\tau_p) \leq 0$$

Objective functions. Using the above constraints and definitions, the LP can optimize (1) total negative slack (TNS)

$$\max: \sum_{\tau_p \in Pins(\tau), \tau \in T} slack(\tau_p)$$

where $Pins(\tau)$ is the set of pins of cell τ, and T is again the set of all sequential elements or endpoints, or (2) worst-negative slack (WNS)

$$\max: WNS$$

where $WNS \leq slack(\tau_p)$ for all pins, or (3) a combination of wirelength and slack

$$\min: \sum_{e \in E} L(e) - \alpha \cdot WNS$$

where E is the set of all nets, α is a constant between 0 and 1 that trades off WNS and wirelength, and $L(e)$ is the HPWL of net e.

8.4 Timing-Driven Routing

In modern ICs, interconnect can contribute substantially to total signal delay. Thus, interconnect delay is a concern during the routing stages. *Timing-driven routing* seeks to minimize one or both of (1) *maximum sink delay*, which is the maximum interconnect delay from the source node to any sink of a given net, and (2) *total wirelength*, which affects the load-dependent delay of the net's driving gate.

For a given signal net *net*, let s_0 be the source node and *sinks* = {s_1, ...,s_n} be the sinks. Let $G = (V,E)$ be a corresponding weighted graph where $V = \{v_0,v_1, \dots ,v_n\}$ represents the source and sink nodes of *net*, and the weight of an edge $e(v_i,v_j) \in E$ represents the routing cost between the terminals v_i and v_j. For any spanning tree T over G, let *radius*(T) be the length of the longest source-sink path in T, and let *cost*(T) be the total edge weight of T.

Because source-sink wirelength reflects source-sink signal delay, i.e., the linear and Elmore delay [8.13] models are well-correlated, a routing tree ideally minimizes both radius and cost. However, for most signal nets, radius and cost cannot be minimized at the same time.

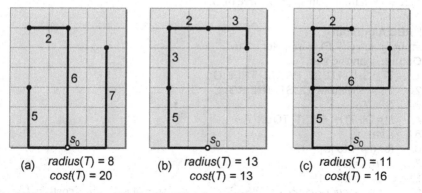

(a) radius(T) = 8
     cost(T) = 20

(b) radius(T) = 13
     cost(T) = 13

(c) radius(T) = 11
     cost(T) = 16

Fig. 8.9 Delay vs. cost, i.e., "shallow" vs. "light", tradeoff in tree construction. (a) A shortest-paths tree. (b) A minimum-cost tree. (c) A compromise with respect to radius (depth) and cost.

Fig. 8.9 illustrates this radius vs. cost ("shallow" vs. "light") tradeoff, where the labels represent edge costs. The tree in Fig. 8.9(a) has minimum radius, and the shortest possible path length from the source to every sink. It is therefore a *shortest-paths tree*, and can be constructed using Dijkstra's algorithm (Sec. 5.6.3). The tree in Fig. 8.9(b) has minimum cost and is a *minimum spanning tree (MST)*, and can be constructed using Prim's algorithm (Sec. 5.6.1). Due to their respective large cost and large radius, neither of these trees may be desirable in practice. The tree in Fig. 8.9(c) is a compromise that has both shallow and light properties.

▶ **8.4.1 The Bounded-Radius, Bounded-Cost Algorithm**

The *bounded-radius, bounded-cost* (*BRBC*) algorithm [8.11] finds a shallow-light spanning tree with provable bounds on both radius and cost. Each of these parameters is within a constant factor of its optimal value. From graph $G(V,E)$, the algorithm first constructs a *subgraph* G' that contains all $v \in V$, and has small cost and small radius. Then, the shortest-paths tree T_{BRBC} in G' will also have small radius and cost because it is a subgraph of G'. T_{BRBC} is determined by a parameter $\varepsilon \geq 0$, which trades off between radius and cost. When $\varepsilon = 0$, T_{BRBC} has minimum radius, and when $\varepsilon = \infty$, T_{BRBC} has minimum cost. More precisely, T_{BRBC} satisfies both

$$radius(T_{BRBC}) \leq (1+\varepsilon) \cdot radius(T_S)$$

where T_S is a shortest-paths tree of G, and

$$cost(T_{BRBC}) \leq \left(1+\frac{2}{\varepsilon}\right) \cdot cost(T_M)$$

where T_M is a minimum spanning tree of G.

BRBC Algorithm
Input: graph $G(V,E)$, parameter $\varepsilon \geq 0$
Output: spanning tree T_{BRBC}
1. T_S = SHORTEST_PATHS_TREE(G)
2. T_M = MINIMUM_COST_TREE(G)
3. $G' = T_M$
4. U = DEPTH_FIRST_TOUR(T_M)
5. $sum = 0$
6.   **for** ($i = 1$ **to** $|U| - 1$)
7. $u_{prev} = U[i]$
8. $u_{curr} = U[i + 1]$
9. $sum = sum + cost_{T_M}[u_{prev}][u_{curr}]$ // sum of $u_{prev} \sim u_{curr}$ costs in T_M
10.   **if** ($sum > \varepsilon \cdot cost_{T_S}[v_0][u_{curr}]$)   // shortest-path cost in T_S
                                                                          //   from source v_0 to u_{curr}
11. G' = ADD(G', PATH(T_S, v_0, u_{curr}))   // add shortest-path edges to G'
12. $sum = 0$ //   and reset sum
13. T_{BRBC} = SHORTEST_PATHS_TREE(G')

To execute the BRBC algorithm, compute a shortest-paths tree T_S of G, and a minimum spanning tree T_M of G (lines 1-2). Initialize the graph G' to T_M (line 3). Let U be the sequence of edges corresponding to any *depth-first tour* of T_M (line 4). This tour traverses each edge of T_M exactly twice (Fig. 8.10), so $cost(U) = 2 \cdot cost(T_M)$. Traverse U while keeping a running total sum of traversed edge costs (lines 6-9). As the traversal visits each node u_{curr}, check whether sum is strictly greater than the cost (distance) between v_0 and u_{curr} in T_S. If so, merge the edges of the $s_0 \sim u_{curr}$ path with

G' and reset *sum* to 0 (lines 11-12). Continue traversing U while repeating this process (lines 6-12). Return T_{BRBC}, a shortest-paths tree over G' (line 13).

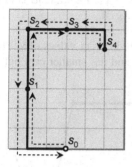

Fig. 8.10 Depth-first tour of the minimum spanning tree in Fig. 8.9 with traversal sequence: $s_0 \rightarrow s_1 \rightarrow s_2 \rightarrow s_3 \rightarrow s_4 \rightarrow s_3 \rightarrow s_2 \rightarrow s_1 \rightarrow s_0$.

▶ **8.4.2 Prim-Dijkstra Tradeoff**

Another method to generate a routing tree trades off radius and cost by using an explicit, quantitative metric. Typically, the minimum cost and minimum radius objectives are optimized by *Prim's minimum spanning tree algorithm* (Sec. 5.6.1) and *Dijkstra's shortest paths tree algorithm* (Sec. 5.6.3), respectively. Although these two algorithms target two different objectives, they construct spanning trees over the set of terminals in very similar ways. From the set of sinks S, each algorithm begins with tree T consisting only of s_0, and iteratively adds a sink s_j and the edge connecting a sink s_i in T to s_j. The algorithms differ only in the *cost function* by which the next sink and edge are chosen.

In Prim's algorithm, sink s_j and edge $e(s_i,s_j)$ are selected to minimize the edge cost between sinks s_i and s_j

$$cost(s_i,s_j)$$

where $s_i \in T$ and $s_j \in S - T$. In Dijkstra's algorithm, sink s_j and edge $e(s_i,s_j)$ are selected to minimize the path cost between source s_0 and sink s_j

$$cost(s_i) + cost(s_i,s_j)$$

where $s_i \in T$, $s_j \in S - T$, and $cost(s_i)$ is the total cost of the shortest path from s_0 to s_i.

To combine the two objectives, the authors of [8.1] proposed the *PD tradeoff*, an explicit compromise between Prim's and Dijkstra's algorithms. This algorithm iteratively adds the sink s_j and the edge $e(s_i,s_j)$ to T such that

$$\gamma \cdot cost(s_i) + cost(s_i,s_j)$$

is minimum over all $s_i \in T$ and $s_j \in S - T$ for a prescribed constant $0 \leq \gamma \leq 1$.

When $\gamma = 0$, the PD tradeoff is identical to Prim's algorithm, and T is a minimum spanning tree T_M. As γ increases, the PD tradeoff constructs spanning trees with progressively higher cost but lower radius. When $\gamma = 1$, the PD tradeoff is identical to Dijkstra's algorithm, and T is a shortest-paths tree T_S.

Fig. 8.11 shows the behavior of the PD tradeoff with different values of γ. The tree in Fig. 8.11(a) is the result of a smaller value of γ, and has smaller cost but larger radius than the tree in Fig. 8.11(b).

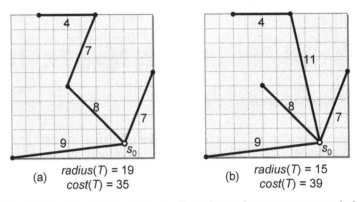

(a) radius(T) = 19
     cost(T) = 35

(b) radius(T) = 15
     cost(T) = 39

Fig. 8.11 Result of the Prim-Dijkstra (PD) tradeoff. Let the cost between any two terminals be their *Manhattan* distance. (a) A tree with $\gamma = 1/4$. (b) A tree with $\gamma = 3/4$, which has higher cost but smaller radius.

▶ 8.4.3 Minimization of Source-to-Sink Delay

The previous subsections described algorithms that seek radius-cost tradeoffs. Since the spanning tree radius reflects actual wire delay, these algorithms indirectly minimize sink delays. However, the wirelength-delay abstraction, as well as the parameters ε in BRBC and γ in the PD tradeoff, prevent direct control of delay. Instead, given a set of sinks S, the *Elmore routing tree* (*ERT*) *algorithm* [8.6] iteratively adds sink s_j and edge $e(s_i,s_j)$ to the growing tree T such that the Elmore delay from the source s_0 to the sink s_j, where $s_i \in T$ and $s_j \in S - T$, is minimized.

Since the ERT algorithm does not treat any particular sink differently from the others, it is classified as a *net-dependent* approach. However, during the actual design and timing optimization, different timing constraints and slacks are imposed for each sink of a multi-pin net.

The sink with the least timing slack is the *critical sink* of the net. A routing tree construction that is oblivious to critical-sink information may create avoidable negative slack, and degrade the overall timing performance of the design. Thus, several routing tree constructions, i.e., *path-dependent approaches*, have been developed that address the *critical-sink routing problem*.

Critical-sink routing tree (CSRT) problem. Given a signal net *net* with source s_0, sinks $S = \{s_1, \ldots, s_n\}$, and sink criticalities $\alpha(i) \geq 0$ for each $s_i \in S$, construct a routing tree T such that

$$\sum_{i=1}^{n} \alpha(i) \cdot t(s_0, s_i)$$

is minimized, where $t(s_0, s_i)$ is the signal delay from source s_0 to sink s_i. The sink criticality $\alpha(i)$ reflects the timing criticality of the corresponding sink s_i. If a sink is on a critical path, then its timing criticality will be greater than that of other sinks.

A *critical-sink Steiner tree* heuristic [8.18] for the CSRT problem [8.6] first constructs a heuristic minimum-cost Steiner tree T_0 over all terminals of S *except* the *critical sink s_c*, the sink with the highest criticality. Then, to reduce $t(s_0, s_c)$, the heuristic adds s_c into T_0 by heuristic variants, e.g., such as the following approaches.

- H_0: introduce a single wire from s_c to s_0.
- H_1: introduce the shortest possible wire that can join s_c to T_0, so long as the path from s_0 to s_c is *monotone*, i.e., of shortest possible total length.
- H_{Best}: try all shortest connections from s_c to edges in T_0, as well as from s_c to s_0. Perform timing analysis on each of these trees and return the one with the lowest delay at s_c.

The time complexity of the critical-sink Steiner heuristic is dominated by the construction of T_0, or by the timing analysis in the H_{Best} variant. Though H_{Best} achieves the best routing solution in terms of timing slack, the other two variants may also provide acceptable combinations of runtime efficiency and solution quality. For high-performance designs, even more comprehensively timing-driven routing tree constructions are needed. Available slack along each source-sink timing arc is best reflected by the *required arrival time (RAT)* at each sink. In the following *RAT tree problem* formulation, each sink of the signal net has a required arrival time which should not be exceeded by the source-sink delay in the routing tree.

RAT tree problem. For a signal net with source s_0 and sink set S, find a minimum-cost routing tree T such that

$$\min_{s \in S} \left(RAT(s) - t(s_0, s) \right) \geq 0$$

Here, $RAT(s)$ is the required arrival time for sink s, and $t(s_0, s)$ is the signal delay in T from source s_0 to sink s. Effective algorithms to solve the RAT tree problem can be found in [8.19]. More information on timing-driven routing can be found in [8.3].

8.5 Physical Synthesis

Recall from Sec. 8.2 that the correct operation of a chip with respect to setup constraints requires that $AAT \leq RAT$ at all nodes. If any nodes violate this condition, i.e., exhibit negative slack, then *physical synthesis*, a collection of timing optimizations, is applied until all slacks are non-negative. There are two aspects to the optimization – *timing budgeting* and *timing correction*. During timing budgeting, target delays are allocated to arcs along timing paths to promote timing closure during the placement and routing stages (Secs. 8.2.2 and 8.3.1), as well as during timing correction (Secs. 8.5.1-8.5.3). During timing correction, the netlist is modified to meet timing constraints using such operations as changing the size of gates, inserting buffers, and netlist restructuring. In practice, a critical path of minimum-slack nodes between two sequential elements is identified, and timing optimizations are applied to improve slack without changing the logical function.

▶ 8.5.1 Gate Sizing

In the standard-cell methodology, each logic gate, e.g., NAND or NOR, is typically available in multiple *sizes* that correspond to different drive strengths. The drive strength is the amount of current that the gate can provide during switching.

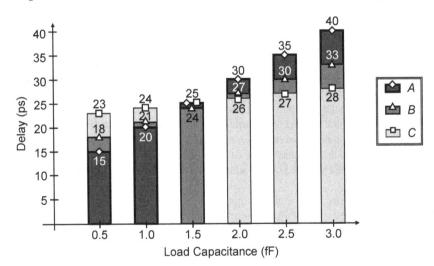

Fig. 8.12 Gate delay vs. load capacitance for three gate sizes A, B, and C, in increasing order.

Fig. 8.12 graphs load capacitance versus delay for versions A, B and C of a gate v, with different sizes (drive strengths), where

$$size(v_A) < size(v_B) < size(v_C)$$

A gate with larger size has lower output resistance and can drive a larger load capacitance with smaller *load-dependent delay*. However, a gate with larger size also has a larger *intrinsic delay* due to the parasitic output capacitance of the gate itself. Thus, when the load capacitance is large,

$$t(v_C) < t(v_B) < t(v_A)$$

because the load-dependent delay dominates. When the load capacitance is small,

$$t(v_A) < t(v_B) < t(v_C)$$

because the intrinsic delay dominates. Increasing $size(v)$ also increases the gate capacitance of v, which, in turn, increases the load capacitance seen by fanin drivers. Although this relationship is not shown, the effects of gate capacitance on the delays of fanin gates will be considered below.

Resizing transformations adjust the size of v to achieve a lower delay (Fig. 8.13). Let $C(p)$ denote the load capacitance of pin p. In Fig. 8.13 (top), the total load capacitance drive by gate v is $C(d) + C(e) + C(f) = 3$ fF. Using gate size A (Fig. 8.13, lower left), the gate delay will be $t(v_A) = 40$ ps, assuming the load-delay relations in Fig. 8.12. However, using gate size C (Fig. 8.13, lower right), the gate delay is $t(v_C) = 28$ ps. Thus, for a load capacitance value of 3 fF, gate delay is improved by 12 ps if v_C is used instead of v_A. Recall that v_C has larger input capacitance at pins a and b, which increases delays of fanin gates. Details of resizing strategies can be found in [8.34]. More information on gate sizing can be found in [8.33].

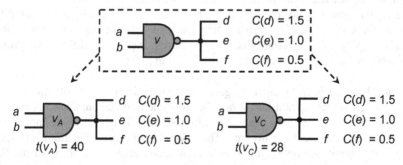

Fig. 8.13 Resizing gate v from gate size A to size C (Fig. 8.12) can achieve a lower gate delay.

▶ 8.5.2 Buffering

A *buffer* is a gate, typically two serially-connected inverters, that regenerates a signal without changing functionality. Buffers can (1) improve timing delays either by speeding up the circuit or by serving as delay elements, and (2) modify transition times to improve signal integrity and coupling-induced delay variation.

In Fig. 8.14 (left), the (actual) arrival time at fanout pins d-h for gate v_B is $t(v_B) =$ 45 ps. Let pins d and e be on the critical path with required arrival times below 35 ps, and let the input pin capacitance of buffer y be 1 fF. Then, adding y reduces the load capacitance of v_B from 5 to 3, and reduces the arrival times at d and e to $t(v_B)$ = 33 ps. That is, the delay of gate v_B is improved by using y to shield v_B from some portion of its initial load capacitance. In Fig. 8.14 (right), after y is inserted, the arrival time at pins f, g and h becomes $t(v_B) + t(y) = 33 + 33 = 66$ ps.

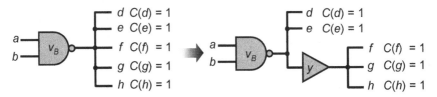

Fig. 8.14 Improving $t(v_B)$ by inserting buffer y to partially shield v_B's load capacitance.

A major drawback of buffering techniques is that they consume the available *area* and increase *power consumption*. Despite the judicious use of buffering by modern tools, the number of buffers has been steadily increasing in large designs due to technology scaling trends, where interconnect is becoming relatively slower compared to gates. In modern high-performance designs, buffers can comprise 10-20% of all standard cell instances, and up to 44% in some designs [8.31].

► **8.5.3 Netlist Restructuring**

Often, the netlist itself can be modified to improve timing. Such changes should not alter the functionality of the circuit, but can use additional gates or modify (rewire) the connections between existing gates to improve driving strength and signal integrity. This section discusses common netlist modifications. More advanced methods for restructuring can be found in [8.25].

Cloning (Replication). Duplicating gates can reduce delay in two situations – (1) when a gate with significant fanout may be slow due to its fanout capacitance, and (2) when a gate's output fans out in two different directions, making it impossible to find a good placement for this gate. The effect on cloning (replication) is to split the driven capacitance between two equivalent gates, at the cost of increasing the fanout of upstream gates.

In Fig. 8.15 (left), using the same load-delay relations of Fig. 8.12, the gate delay $t(v_B)$ of gate v_B is 45 ps. However, In Fig. 8.15 (right), after cloning, $t(v_A) = 30$ ps and $t(v_B) = 33$ ps. Cloning also increases the input pin capacitance seen by the fanin gates that generate signals a and b. In general, cloning allows more freedom for local placement, e.g., the instance v_A can be placed close to sinks d and e, while the instance v_B can be placed close to sinks f, g and h, with the tradeoff of increased congestion and routing cost.

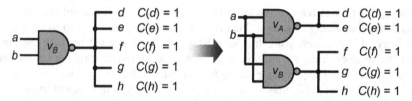

Fig. 8.15 Cloning or duplicating gates to reduce maximum local fanout.

When the downstream capacitance is large, buffering may be a better alternative than cloning because buffers do not increase the fanout capacitance of upstream gates. However, buffering cannot replace placement-driven cloning. An exercise at the end of this chapter expands further upon this concept.

The second application of cloning allows the designers to replicate gates and place each clone closer to its downstream logic. In Fig. 8.16, v drives five signals d-h, where signals d, e and f are close, and g and h are located much farther away. To mitigate the large fanout of v and the large interconnect delay caused by remote signals, gate v is cloned. The original gate v remains with only signals d, e, and f, and a new copy of v (v') is placed closer to g and h.

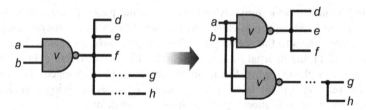

Fig. 8.16 Cloning transformation: a driving gate is duplicated to reduce remoteness of its fanouts.

Redesign of fanin tree. The logic design phase often provides a circuit with the minimum number of *logic levels*. Minimizing the maximum number of gates on a path between sequential elements tends to produce a *balanced* circuit with similar path delays from inputs to outputs. However, input signals may arrive at varied times, so the minimum-level circuit may not be timing-optimal. In Fig. 8.17, the arrival time $AAT(f)$ of pin f is 6 no matter how the input signals are mapped to gate input pins. However, the unbalanced network has a shorter input-output path which can be used by a later-arriving signal, where $AAT(f) = 5$.

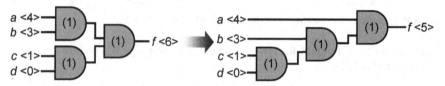

Fig. 8.17 Redesigning a fanin tree to have smaller input-to-output delay. The arrival times are denoted in angular brackets, and the delay are denoted in parentheses.

Redesign of fanout tree. In the same spirit as Fig. 8.17, it is possible to improve timing by rebalancing the output load capacitance in a fanout tree so as to reduce the delay of the longest path. In Fig. 8.18, buffer y_1 is needed because the load capacitance of critical path $path_1$ is large. However, by redesigning the fanout tree to reduce the load capacitance of $path_1$, use of the buffer y_1 can be avoided. Increased delay on $path_2$ may be acceptable if that path is not critical even after the load capacitance of buffer y_2 is increased.

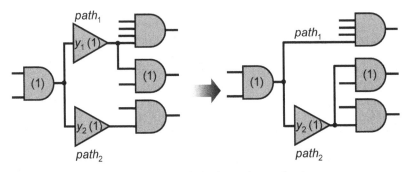

Fig. 8.18 Redesign of a fanout tree to reduce the load capacitance of $path_1$.

Swapping commutative pins. Although the input pins of, e.g., a two-input NAND gate are logically equivalent, in the actual transistor network they will have different delays to the output pin. When the *pin node convention* is used for STA (Sec. 8.2.1), the internal input-output arcs will have different delays. Hence, path delays can change when the input pin assignment is changed. The rule of thumb for pin assignment is to assign a later- (sooner-) arriving signal to an equivalent input pin with shorter (longer) input-output delay.

In Fig. 8.19, the internal timing arcs are labeled with corresponding delays in parentheses, and pins a, b, c and f are labeled with corresponding arrival times in angular brackets. In the circuit on the left, the arrival time at f can be improved from 5 to 3 by swapping pins a and c.

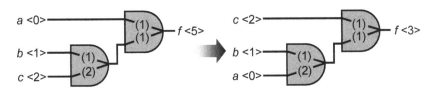

Fig. 8.19 Swapping commutative pins to reduce the arrival time at f.

More advanced techniques for pin assignment and swapping of commutative pins can be found in [8.9].

Gate decomposition. In CMOS designs, a gate with multiple inputs usually has larger size and capacitance, as well as a more complex transistor-level network topology that is less efficient with respect to speed metrics such as *logical effort* [8.32]. Decomposition of multiple-input gates into smaller, more efficient gates can decrease delay and capacitance while retaining the same Boolean functionality. Fig. 8.20 illustrates the decomposition of a multiple-input gate into equivalent networks of two- and three-input gates.

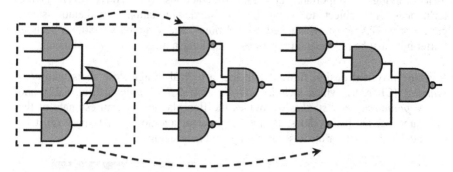

Fig. 8.20 Gate decomposition of a complex network into alternative networks.

Boolean restructuring. In digital circuits, Boolean logic can be implemented in multiple ways. In the example of Fig. 8.21, $f(a,b,c) = (a + b)(a + c) \equiv a + bc$ (distributive law) can be exploited to improve timing when two functions have overlapping logic or share logic nodes. The figure shows two functions $x = a + bc$ and $y = ab + c$ with arrival times $AAT(a) = 4$, $AAT(b) = 1$, and $AAT(c) = 2$. When implemented using a common node $a + c$, the arrival times of x and y are $AAT(x) = AAT(y) = 6$. However, implementing x and y separately achieves $AAT(x) = 5$ and $AAT(y) = 6$.

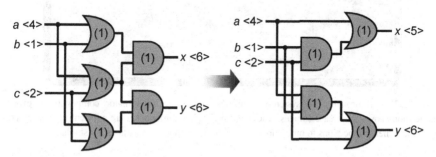

Fig. 8.21 Restructuring using logic properties, e.g., the distributive law, to improve timing.

Reverse transformations. Timing optimizations such as buffering, sizing, and cloning increase the original area of the design. This change can cause the design to be *illegal*, as some new cells can now overlap with others. To maintain legality, either (1) perform the respective reverse operations *unbuffering*, *downsizing*, and *merging*, or (2) perform *placement legalization* after all timing corrections.

8.6 Performance-Driven Design Flow

The previous sections have presented several algorithms and techniques to improve timing of digital circuit designs. This section combines all these optimizations in a consistent *performance-driven physical design flow*, which seeks to satisfy timing constraints, i.e., "close on timing". Due to the nature of performance optimizations, their ordering is important, and their interactions with conventional layout techniques are subject to a number of subtle limitations. Evaluation steps, particularly STA, must be invoked several times, and some optimizations, such as buffering, must be redone multiple times to facilitate a more accurate evaluation.

Baseline physical design flow. Recall that a typical design flow starts with *chip planning* (Chap. 3), which includes *I/O placement, floorplanning* (Fig. 8.22), and *power planning. Trial synthesis* provides the floorplanner with an estimate of the total area needed by modules. Besides logic area, additional whitespace must be allocated to account for buffers, routability, and gate sizing.

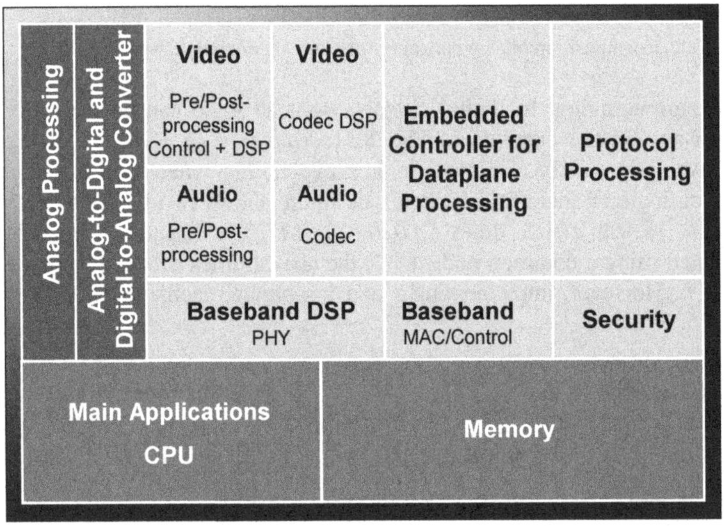

Fig. 8.22 A floorplan of a system-on-chip (SoC) design. Each major component is given dimensions based on area estimates. The audio and video components are adjacent to each other, given that their connections to other blocks and their performance constraints are similar.

Then, *logic synthesis* and *technology mapping* produce a gate- (cell)-level netlist from a high-level specification, which is tailored to a specific technology library. Next, *global placement* assigns locations to each movable object (Chap. 4). As illustrated in Fig. 8.23, most of the cells are clustered in highly concentrated regions (colored black). As the iterations progress, the cells are gradually spread across the chip, such that they no longer overlap (colored light gray).

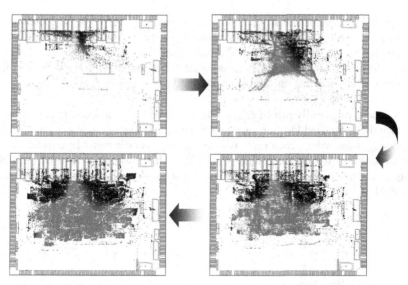

Fig. 8.23 The progression of cell spreading during global placement in a large, flat (non-floorplanned) ASIC design with fixed macro blocks. Darker shades indicate greater cell overlap while lighter shades indicate smaller cell overlap.

These locations, however, do not have to be aligned with cell rows or sites, and can allow slight cell overlap. To ensure that the overlap is small, it is common to (1) establish a uniform grid, (2) compute the total area of objects in each grid square, and (3) limit this total by the area available in the square.

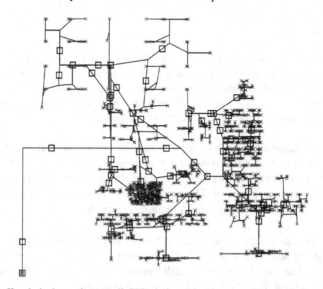

Fig. 8.24 Buffered clock tree in a small CPU design. The clock source is in the lower left corner. Crosses (×) indicate sinks, and boxes (□) indicate buffers. Each diagonal segment represents a horizontal plus a vertical wire (L-shape), the choice of which can be based on routing congestion.

After global placement, the sequential elements are legalized. Once the locations of sequential elements are known, a *clock network* (Chap. 7) is generated. ASICs, SoCs and low-power (mobile) CPUs commonly use clock trees (Fig. 8.24), while high-performance microprocessors incorporate structured and hand-optimized clock distribution networks that may combine trees and meshes [8.30][8.31].

The locations of globally placed cells are first temporarily rounded to a uniform grid, and then these rounded locations are connected during *global routing* (Chap. 5) and *layer assignment*, where each route is assigned to a specific metal layer. The routes indicate areas of wiring congestion (Fig. 8.25). This information is used to guide *congestion-driven detailed placement* and *legalization* of combinational elements (Chap. 4).

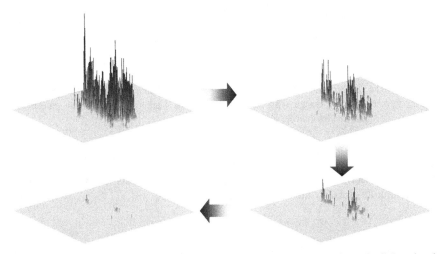

Fig. 8.25 Progression of congestion maps through iterations of global routing. The light-colored areas are those that do not have congestion; dark-colored peaks indicate congested regions. Initially, several dense clusters of wires create edges that are far over capacity. After iterations of rip-up and reroute, the route topologies are changed, alleviating the most congested areas. Though more regions can become congested, the maximum congestion is reduced.

While detailed placement conventionally comes before global routing, the reverse order can reduce overall congestion and wirelength [8.28]. Note that EDA flows require a legal placement before global routing. In this case, legalization will be performed after global placement. The global routes of signal nets are then assigned to physical routing tracks during *detailed routing* (Chap. 6).

The layout generated during the place-and-route stage is subjected to *reliability*, *manufacturability* and *electrical verification*. During *mask generation*, each standard cell and each route are represented by collections of rectangles in a format suitable for generating optical lithography masks for chip fabrication.

This baseline PD flow is illustrated in Fig. 8.26 with white boxes.

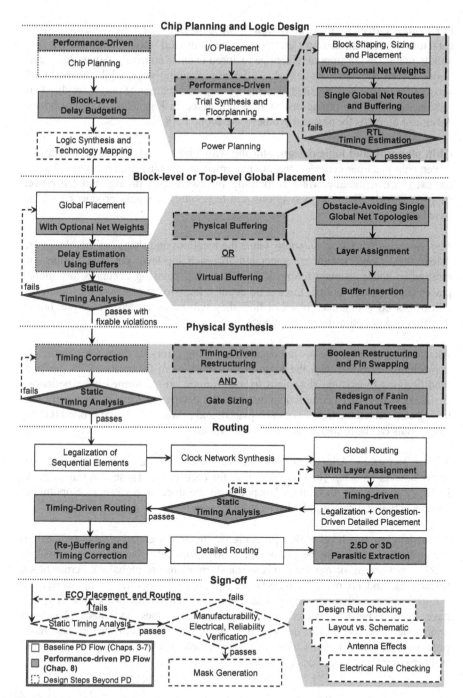

Fig. 8.26 Integrating optimizations covered in Chaps. 3-8 into a performance-driven design flow. Some tools bundle several optimization steps, which changes the appearance of the flow to users and often alters the user interface. Alternatives to this flow are discussed in this section.

Performance-driven physical design flow. Extending the baseline design flow, contemporary industrial flows are typically built around static timing analysis and seek to minimize the amount of change required to close on timing. Some flows start timing-driven optimizations as early as the chip planning stage, while others do not account for timing until detailed placement to ensure accuracy of timing results. This section discusses the timing-driven flow illustrated in Fig. 8.26 with gray boxes. Advanced methods for physical synthesis are found in [8.4].

Chip planning and logic design. Starting with a high-level design, performance-driven chip planning generates the I/O placement of the pins and rectangular blocks for each circuit module while accounting for block-level timing, and the power supply network. Then, logic synthesis and technology mapping produces a netlist based on delay budgets.

Performance-driven chip planning. Once the locations and shapes of the blocks are determined, global routes are generated for each top-level net, and buffers are inserted to better estimate timing [8.2]. Since chip planning occurs before global placement or global routing, there is no detailed knowledge of where the logic cells will be placed within each block or how they will be connected. Therefore, buffer insertion makes optimistic assumptions.

After buffering, STA checks the design for timing errors. If there are a sufficient number of violations, then the logic blocks must be re-floorplanned. In practice, modifications to existing floorplans to meet timing are performed by experienced designers with little to no automation. Once the design has satisfied or mostly met timing constraints, the I/O pins can be placed, and power (VDD) and ground (GND) supply rails can be routed around floorplan blocks.

Timing budgeting. After performance-driven floorplanning, delay budgeting sets upper bounds on *setup* (long path) timing for each block. These constraints guide logic synthesis and technology mapping to produce a performance-optimized gate-level netlist, using standard cells from a given library.

Block-level or top-level global placement. Starting at global placement, timing-driven optimizations can be performed at the *block level*, where each individual block is optimized, or *top level*, where transformations are global, i.e., cross block boundaries, and all movable objects are optimized.[4] Block-level approaches are useful for designs that have many macro blocks or *intellectual properties* (*IPs*) that have already been optimized and have specific shapes and sizes. Top-level approaches are useful for designs that have more freedom or do not reuse previously-designed logic; a hierarchical methodology offers more parallelism and is more common for large design teams.

---

[4] In hierarchical design flows, different designers concurrently perform top-level placement and block-level placement.

Buffer insertion. To better estimate and improve timing, buffers are inserted to break any extremely long or high fanout nets (Sec. 8.5.2). This can be done either *physically,* where buffers are directly added to the placement, or *virtually,* where the impact of buffering is included in delay models, but the netlist is not modified.

Physical buffering. Physical buffering [8.1] performs the full process of buffer insertion by (1) generating obstacle-avoiding global net topologies for each net, (2) estimating which metal layers the route uses, and (3) actually inserting buffers (Fig. 8.27).

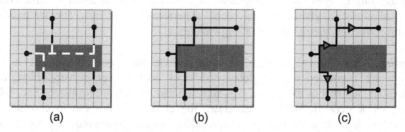

(a)   (b)   (c)

Fig. 8.27 Physical buffering for timing estimation. (a) A five-pin net is routed with a minimum Steiner tree topology that does not avoid a routing obstacle (shown in gray). (b) The net routed with an obstacle-avoiding Steiner tree topology. (c) The buffered topology offers a relatively accurate delay estimation.

Virtual buffering [8.24], on the other hand, estimates the delay by modeling every pin-to-pin connection as an optimally buffered line with linear delay [8.23] as

$$t_{LD}(net) = L(net) \cdot \left(R(B) \cdot C(w) + R(w) \cdot C(B) + \sqrt{2 \cdot R(B) \cdot C(B) \cdot R(w) \cdot C(w)} \right)$$

where *net* is the net, $L(net)$ is the total length of *net*, $R(B)$ and $C(B)$ are the respective intrinsic resistance and capacitance of the buffer, and $R(w)$ and $C(w)$ are the respective unit wire resistance and capacitance. Though no buffers are added to the netlist, they are assumed for timing purposes. When timing information becomes more accurate, subsequent re-buffering steps often remove any existing buffers and re-insert them from scratch. In this context, virtual buffering saves effort, while preserving the accuracy of timing analysis. Physical buffering can avoid unnecessary upsizing of drivers and is more accurate than virtual buffering, but also more time-consuming.

Once buffering is complete, the design is checked for timing violations using static timing analysis (Sec. 8.2.1). Unless timing is met, the design returns to buffering, global placement, or, in some cases, to logic synthesis. When timing constraints are mostly met, the design moves on to timing correction, which includes gate sizing (Sec. 8.5.1) and timing-driven netlist restructuring (Sec. 8.5.3). Subsequently, another timing check is performed using STA.

Physical synthesis. After buffer insertion, physical synthesis applies several timing correction techniques (Sec. 8.5) such as operations that modify the pin ordering or the netlist at the gate level, to improve delay on critical paths.

Timing correction. Methods such as *gate sizing* increase (decrease) the size of a physical gate to speed up (slow down) the circuit. Other techniques such as *redesign of fanin and fanout trees*, *cloning*, and *pin swapping* reduce timing by rebalancing existing logic to reduce load capacitance for timing-critical nets. Transformations such as *gate decomposition* and *Boolean restructuring* modify logic locally to improve timing by merging or splitting logic nodes from different signals. After physical synthesis, another timing check is performed. If it fails, another pass of timing correction attempts to fix timing violations.

Routing. After physical synthesis, all combinational and sequential elements in the design are connected during global and clock routing, respectively. First, the sequential elements of the design, e.g., flip-flop and latches, are legalized (Sec. 4.4). Then, clock network synthesis generates the clock tree or mesh to connect all sequential elements to the clock source. Modern clock networks require a number of large clock buffers;[5] performing clock-network design *before* detailed placement allows these buffers to be placed appropriately. Given the clock network, the design can be checked for *hold-time (short path) constraints*, since the clock skews are now known, whereas only *setup (long path) constraints* could be checked before.

Layer assignment. After clock-network synthesis, global routing assigns global route topologies to connect the combinational elements. Then, layer assignment matches each global route to a specific metal layer. This step improves the accuracy of delay estimation because it allows the use of appropriate *resistance-capacitance* (*RC*) parasitics for each net. Note that clock routing is performed before signal-net routing when the two share the same metal layers – clock routes take precedence and should not detour around signal nets.

Timing-driven detailed placement. The results of global routing and layer assignment provide accurate estimates of wire congestion, which is then used by a congestion-driven detailed placer [8.10][8.35]. The cells are (1) spread to remove overlap among objects and decrease routing congestion, (2) snapped to standard-cell rows and legal cell sites, and then (3) optimized by swaps, shifts and other local changes. To incorporate timing optimizations, *either* perform (1) non-timing-driven legalization followed by timing-driven detailed placement, *or* (2) perform timing-driven legalization followed by non-timing-driven detailed placement. After detailed placement, another timing check is performed. If timing fails, the design could be globally re-routed or, in severe cases, globally re-placed.

To give higher priority to the clock network, the sequential elements can be legalized first, and then followed by global and detailed routing. With this approach,

---

[5] These buffers are legalized immediately when added to the clock network.

signal nets must route around the clock network. This is advantageous for large-scale designs, as clock trees are increasingly becoming a performance bottleneck. A variant flow, such as the industrial flow described in [8.28], first fully legalizes the locations of all cells, and then performs detailed placement to recover wirelength.

Another variant performs detailed placement before clock network synthesis, and then is followed by legalization and several optimization steps.[6] After the clock network has been synthesized, another pass of setup optimization is performed. Hold violations may be addressed at this time or, optionally, after routing and initial STA.

Timing-driven routing. After detailed placement, clock network synthesis and post-clock network optimization, the *timing-driven routing* phase aims to fix the remaining timing violations. Algorithms discussed in Sec. 8.4 include generating *minimum-cost, minimum-radius trees* for critical nets (Secs. 8.4.1-8.4.2), and *minimizing the source-to-sink delay* of critical sinks (Sec. 8.4.3).

If there are still outstanding timing violations, further optimizations such as re-buffering and late timing corrections are applied. An alternative is to have designers manually tune or fix the design by relaxing some design constraints, using additional logic libraries, or exploiting design structure neglected by automated tools. After this time-consuming process, another timing check is performed. If timing is met, then the design is sent to detailed routing, where each signal net is assigned to specific routing tracks. Typically, incremental STA-driven *Engineering Change Orders* (*ECOs*) are applied to fix timing violations after detailed placement; this is followed by ECO placement and routing. Then, *2.5D* or *3D parasitic extraction* determines the electromagnetic impact on timing based on the routes' shapes and lengths, and other technology-dependent parameters.

Signoff. The last few steps of the design flow validate the layout and timing, as well as fix any outstanding errors. If a timing check fails, ECO minimally modifies the placement and routing such that the violation is fixed and no new errors are introduced. Since the changes made are very local, the algorithms for ECO placement and ECO routing differ from the traditional place and route techniques discussed in Chaps. 4-7.

After completing timing closure, *manufacturability*, *reliability* and *electrical verification* ensure that the design can be successfully fabricated and will function correctly under various environmental conditions. The four main components are equally important and can be performed in parallel to improve runtime.

— *Design Rule Checking* (*DRC*) ensures that the placed-and-routed layout meets all technology-specified design rules e.g., minimum wire spacing and width.

---

[6] These include post-clock-network-synthesis optimizations, post-global-routing optimizations, and post-detailed-routing optimizations.

- *Layout vs. Schematic* (*LVS*) checking ensures the placed-and-routed layout matches the original netlist.
- *Antenna Checks* seek to detect undesirable *antenna effects*, which may damage a transistor during plasma-etching steps of manufacturing by collecting excess charge on metal wires that are connected to PN-junction nodes. This can occur when a route consists of multiple metal layers and a charge is induced on a metal layer during fabrication.
- *Electric Rule Checking* (*ERC*) finds all potentially dangerous electric connections, such as floating inputs and shorted outputs.

Once the design has been physically verified, *optical-lithography masks* are generated for manufacturing.

8.7   8.7 Conclusions

This chapter explained how to combine timing optimizations into a comprehensive physical design flow. In practice, the flow described in Sec. 8.6 (Fig. 8.26) can be modified based on several factors, including

- Design type.
    - ASIC, microprocessor, IP, analog, mixed-mode.
    - Datapath-heavy specifications may require specialized tools for structured placement or manual placement. Datapaths typically have shorter wires and require fewer buffers for high-performance layout.
- Design objectives.
    - High-performance, low-power or low-cost.
    - Some high-performance optimizations, such as buffering and gate sizing, increase circuit area, thus increasing circuit power and chip cost.
- Additional optimizations.
    - *Retiming* shifts locations of registers among combinational gates to better balance delay.
    - *Useful skew* scheduling, where the clock signal arrives earlier at some flip-flops and later at others, to adjust timing.
    - *Adaptive body-biasing* can improve the leakage current of transistors.
- Additional analyses.
    - Multi-corner and multi-mode static timing analysis, as industrial ASICs and microprocessors are often optimized to operate under different temperatures and supply voltages.
    - Thermal analysis is required for high-performance CPUs.
- Technology node, typically specified by the *minimum feature size*.
    - Nodes < 180 nm require timing-driven placement and routing flows, as lumped-capacitance models are inadequate for performance estimation.

- Nodes < 130 nm require timing analysis with signal integrity, i.e., interconnect coupling capacitances and the resulting delay increase (decrease) of a given *victim net* when a neighboring *aggressor* net switches simultaneously in the opposite (same) direction.
- Nodes < 90 nm require additional *resolution enhancement techniques* (*RET*) for lithography.
- Nodes < 65 nm require power-integrity (e.g., IR drop-aware timing, electromigration reliability) analysis flows.
- Nodes < 45 nm require additional statistical power-performance tradeoffs at the transistor level.
- Nodes < 32 nm impose significant limitations on detailed routing, known as *restricted design rules* (*RDRs*), to ensure manufacturability.
- Available tools.
  - In-house software, commercial EDA tools [8.34].
- Design size and the extent of design reuse.
  - Larger designs often include more global interconnect, which may become a performance bottleneck and typically requires buffering.
  - IP blocks are typically represented by hard blocks during floorplanning.
- Design team size, required time-to-market, available computing resources.
  - To shorten time-to-market, one can leverage a large design team by partitioning the design into blocks and assigning blocks to teams.
  - After floorplanning, each block can be laid out in parallel; however, *flat* optimization (no partitioning) sometimes produces better results.

Reconfigurable fabrics such as FPGAs require less attention to buffering, due to already-buffered programmable interconnect. Wire congestion is often negligible for FPGAs because interconnect resources are overprovisioned. However, FPGA detailed placement must satisfy a greater number of constraints than placement for other circuit types, and global routing must select from a greater variety of interconnect types. Electrical and manufacturability checks are unnecessary for FPGAs, but technology mapping is more challenging, as it affects the area and timing to a greater extent, and can benefit more from the use of physical information. Therefore, modern physical-synthesis flows for FPGAs perform global placement, often in a trial mode, between logic synthesis and technology mapping.

Physical design flows will require additional sophistication to support increasing transistor densities in semiconductor chips. The advent of future technology nodes – 28 nm, 22 nm and 16 nm – will bring into consideration new electrical and manufacturing-related phenomena, while increasing uncertainty in device parameters [8.22]. Further increase in transistor counts may require integrating multiple chips into three-dimensional integrated circuits, thus changing the geometry of fundamental physical design optimizations [8.36]. Nevertheless, the core optimizations described in this chapter will remain vital in chip design.

Chapter 8 Exercises

Exercise 1: Static Timing Analysis

Given the logic circuit below, draw the timing graph (a), and determine the (b) AAT, (c) RAT, and (d) slack of each node. The AATs of the inputs are in angular brackets, the delays are in parentheses, and the RAT of the output is in square brackets.

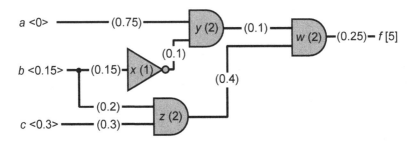

Exercise 2: Timing-Driven Routing

Given the terminal locations of a signal net, assume that all distances are Manhattan, and that no Steiner point is used when routing. Construct the spanning tree T, and calculate $radius(T)$ and $cost(T)$, for each of the following.

(a) Prim-Dijkstra tradeoff with $\gamma = 0$ (Prim's MST algorithm).
(b) Prim-Dijkstra tradeoff with $\gamma = 1$ (Dijkstra's algorithm).
(c) Prim-Dijkstra tradeoff with $\gamma = 0.5$.

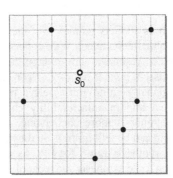

Exercise 3: Buffer Insertion for Timing Improvement

For the logic circuit and load capacitances on the following page, assume that available gate sizes and timing performances are similar to those in Fig. 8.12. Assume that gate delay always increases linearly with load capacitance. Let the input capacitance of buffer y be 0.5 fF, 1 fF, and 2 fF with sizes A, B and C, respectively. Determine the size for buffer y that minimizes the AAT of sink c.

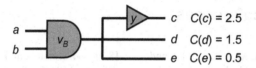

c $C(c) = 2.5$
d $C(d) = 1.5$
e $C(e) = 0.5$

Exercise 4: Timing Optimization

List at least two timing optimizations covered *only* in this chapter (not mentioned beforehand). Describe these optimizations in your own words and discuss scenarios in which (1) they can be useful and (2) they can be harmful.

Exercise 5: Cloning vs. Buffering

List and explain scenarios where cloning results in better timing improvements than buffering, and vice-versa. Explain why both methods are necessary for timing-driven physical synthesis.

Exercise 6: Physical Synthesis

In terms of timing corrections such as buffering, gate sizing, and cloning, when are their reverse transformations useful? In what situations will a given timing correction cause the design to be illegal? Explain for each timing correction.

Chapter 8 References

[8.1] C. J. Alpert, M. Hrkic, J. Hu and S. T. Quay, "Fast and Flexible Buffer Trees that Navigate the Physical Layout Environment", *Proc. Design Autom. Conf.*, 2004, pp. 24-29.

[8.2] C. J. Alpert, J. Hu, S. S. Sapatnekar and C. N. Sze, "Accurate Estimation of Global Buffer Delay Within a Floorplan", *IEEE Trans. on CAD* 25(6) (2006), pp. 1140-1146.

[8.3] C. J. Alpert, T. C. Hu, J. H. Huang and A. B. Kahng, "A Direct Combination of the Prim and Dijkstra Constructions for Improved Performance-Driven Global Routing", *Proc. Intl. Conf. on Circuits and Sys.*, 1993, pp. 1869-1872.

[8.4] C. J. Alpert, S. K. Karandikar, Z. Li, G.-J. Nam, S. T. Quay, H. Ren, C. N. Sze, P. G. Villarrubia and M. C. Yildiz, "Techniques for Fast Physical Synthesis", *Proc. IEEE* 95(3) (2007), pp. 573-599.

[8.5] C. J. Alpert, D. P. Mehta and S. S. Sapatnekar, eds., *Handbook of Algorithms for Physical Design Automation*, CRC Press, 2009.

[8.6] K. D. Boese, A. B. Kahng and G. Robins, "High-Performance Routing Trees with Identified Critical Sinks", *Proc. Design Autom. Conf.*, 1993, pp. 182-187.

[8.7] M. Burstein and M. N. Youssef, "Timing Influenced Layout Design", *Proc. Design Autom. Conf.*, 1985, pp. 124-130.

[8.8] T. F. Chan, J. Cong and E. Radke, "A Rigorous Framework for Convergent Net Weighting Schemes in Timing-Driven Placement", *Proc. Intl Conf. on CAD*, 2009, pp. 288-294.

[8.9] K.-H. Chang, I. L. Markov and V. Bertacco, "Postplacement Rewiring by Exhaustive Search for Functional Symmetries", *ACM Trans. on Design Autom. of Electronic Sys.* 12(3) (2007), pp. 1-21.

[8.10] A. Chowdhary, K. Rajagopal, S. Venkatesan, T. Cao, V. Tiourin, Y. Parasuram and B. Halpin, "How Accurately Can We Model Timing in a Placement Engine?", *Proc. Design Autom. Conf.*, 2005, pp. 801-806.

[8.11] J. Cong, A. B. Kahng, G. Robins, M. Sarrafzadeh and C. K. Wong, "Provably Good Algorithms for Performance-Driven Global Routing", *Proc. Intl. Symp. on Circuits and Sys.*, 1992, pp. 2240-2243.

[8.12] A. E. Dunlop, V. D. Agrawal, D. N. Deutsch, M. F. Jukl, P. Kozak and M. Wiesel, "Chip Layout Optimization Using Critical Path Weighting", *Proc. Design Autom. Conf.*, 1984, pp. 133-136.

[8.13] W. C. Elmore, "The Transient Response of Damped Linear Networks with Particular Regard to Wideband Amplifiers", *J. Applied Physics* 19(1) (1948), pp. 55-63.

[8.14] H. Eisenmann and F. M. Johannes, "Generic Global Placement and Floorplanning", *Proc. Design Autom. Conf.*, 1998, pp. 269-274.

[8.15] S. Ghiasi, E. Bozorgzadeh, P.-K. Huang, R. Jafari and M. Sarrafzadeh, "A Unified Theory of Timing Budget Management", *IEEE Trans. on CAD* 25(11) (2006), pp. 2364-2375.

[8.16] B. Halpin, C. Y. R. Chen and N. Sehgal, "Timing Driven Placement Using Physical Net Constraints", *Proc. Design Autom. Conf.*, 2001, pp. 780-783.

[8.17] P. S. Hauge, R. Nair and E. J. Yoffa, "Circuit Placement for Predictable Performance", *Proc. Intl. Conf. on CAD*, 1987, pp. 88-91.

[8.18] T. I. Kirkpatrick and N. R. Clark, "PERT as an Aid to Logic Design", *IBM J. Research and Development* 10(2) (1966), pp. 135-141.

[8.19] J. Lillis, C.-K. Cheng, T.-T. Y. Lin and C.-Y. Ho, "New Performance Driven Routing Techniques With Explicit Area/Delay Tradeoff and Simultaneous Wire Sizing", *Proc. Design Autom. Conf.*, 1996, pp. 395-400.

[8.20] R. Nair, C. L. Berman, P. S. Hauge and E. J. Yoffa, "Generation of Performance Constraints for Layout", *IEEE Trans. on CAD* 8(8) (1989), pp. 860-874.

[8.21] M. Orshansky and K. Keutzer, "A General Probabilistic Framework for Worst Case Timing Analysis", *Proc. Design Autom. Conf.*, 2002, pp. 556-561.

[8.22] M. Orshansky, S. Nassif and D. Boning, *Design for Manufacturability and Statistical Design: A Constructive Approach*, Springer, 2008.

[8.23] R. H. J. M. Otten and R. K. Brayton, "Planning for Performance", *Proc. Design Autom. Conf.*, 1998, pp. 122-127.

[8.24] D. A. Papa, T. Luo, M. D. Moffitt, C. N. Sze, Z. Li, G.-J. Nam, C. J. Alpert and I. L. Markov, "RUMBLE: An Incremental, Timing-Driven, Physical-Synthesis Optimization Algorithm", *IEEE Trans. on CAD* 27(12) (2008), pp. 2156-2168.

[8.25] S. M. Plaza, I. L. Markov and V. Bertacco, "Optimizing Nonmonotonic Interconnect Using Functional Simulation and Logic Restructuring", *IEEE Trans. on CAD* 27(12) (2008), pp. 2107-2119.

[8.26] K. Rajagopal, T. Shaked, Y. Parasuram, T. Cao, A. Chowdhary and B. Halpin, "Timing Driven Force Directed Placement with Physical Net Constraints", *Proc. Intl. Symp. on Phys. Design*, 2003, pp. 60-66.

[8.27] H. Ren, D. Z. Pan and D. S. Kung, "Sensitivity Guided Net Weighting for Placement Driven Synthesis", *IEEE Trans. on CAD* 24(5) (2005) pp. 711-721.

[8.28] J. A. Roy, N. Viswanathan, G.-J. Nam, C. J. Alpert and I. L. Markov, "CRISP: Congestion Reduction by Iterated Spreading During Placement", *Proc. Intl. Conf. on CAD*, 2009, pp. 357-362.

[8.29] M. Sarrafzadeh, M. Wang and X. Yang, *Modern Placement Techniques*, Kluwer, 2003.

[8.30] R. S. Shelar, "Routing With Constraints for Post-Grid Clock Distribution in Microprocessors", *IEEE Trans. on CAD* 29(2) (2010), pp. 245-249.

[8.31] R. S. Shelar and M. Patyra, "Impact of Local Interconnects on Timing and Power in a High Performance Microprocessor", *Proc. Intl. Symp. on Phys. Design*, 2010, pp. 145-152.

[8.32] I. Sutherland, R. F. Sproull and D. Harris, *Logical Effort: Designing Fast CMOS Circuits*, Morgan Kaufmann, 1999.

[8.33] H. Tennakoon and C. Sechen, "Nonconvex Gate Delay Modeling and Delay Optimization", *IEEE Trans. on CAD* 27(9) (2008), pp. 1583-1594.

[8.34] M. Vujkovic, D. Wadkins, B. Swartz and C. Sechen, "Efficient Timing Closure Without Timing Driven Placement and Routing", *Proc. Design Autom. Conf.*, 2004, pp. 268-273.

[8.35] J. Westra and P. Groeneveld, "Is Probabilistic Congestion Estimation Worthwhile?", *Proc. Sys. Level Interconnect Prediction*, 2005, pp. 99-106.

[8.36] Y. Xie, J. Cong and S. Sapatnekar, eds., *Three-Dimensional Integrated Circuit Design: EDA, Design and Microarchitectures*, Springer, 2010.

Appendix A

Solutions to Chapter Exercises

A   Solutions to Chapter Exercises

Chapter 2: Netlist and System Partitioning

Exercise 1: KL Algorithm

Maximum gains and the swaps of nodes with positive gain per iteration i are given.

$i = 1: \Delta g_1 = D(a) + D(f) - 2c(a,f) = 2 + 1 - 0 = 3 \rightarrow$ swap nodes a and f, $G_1 = 3$.
$i = 2: \Delta g_2 = D(c) + D(e) - 2c(c,e) = -2 + 0 - 0 = -2 \rightarrow$ swap nodes c and e, $G_2 = 1$.
$i = 3: \Delta g_3 = D(b) + D(d) - 2c(b,d) = -1 + 0 - 0 = -1 \rightarrow$ swap nodes b and d, $G_3 = 0$.

Maximum gain is $G_m = 3$ with $m = 1$.
Therefore, swap nodes a and f.
Graph after Pass 1 (right).

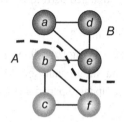

Exercise 2: Critical Nets and Gain During the FM Algorithm

(a)   Table recording (1) the cell moved, (2) critical nets before the move, (3) critical nets after the move, and (4) which cells require a gain update.

Cell moved	Critical nets before the move	Critical nets after the move	Which cells require a gain update
a	--	N_1	b
b	--	N_1	a
c	--	N_1	d
d	N_3	N_1,N_3	c,h,i
e	N_2	N_2	f,g
f	N_2	N_2	e,g
g	N_2	N_2	e,f
h	N_3	N_3	d,i
i	N_3	N_3	d,h

(b) Gains for each cell.

$\Delta g_1(a) = 0$ $\Delta g_1(b) = 0$ $\Delta g_1(c) = 0$
$\Delta g_1(d) = 0$ $\Delta g_1(e) = -1$ $\Delta g_1(f) = -1$
$\Delta g_1(g) = -1$ $\Delta g_1(h) = 1$ $\Delta g_1(i) = 0$

Exercise 3: FM Algorithm

Pass 2, iteration $i = 1$

Gain values: $\Delta g_1(a) = -1, \Delta g_1(b) = -1, \Delta g_1(c) = 0, \Delta g_1(d) = 0, \Delta g_1(e) = -1.$

Cells c and d have maximum gain value $\Delta g_1 = 0$.

Balance criterion after moving cell c: $area(A) = 4$.

Balance criterion after moving cell d: $area(A) = 1$.

Cell c meets the balance criterion better.

Move cell c, updated partitions: $A_1 = \{d\}$, $B_1 = \{a,b,c,e\}$, with fixed cells $\{c\}$.

Pass 2, iteration $i = 2$

Gain values: $\Delta g_2(a) = -2, \Delta g_2(b) = -2, \Delta g_2(d) = 2, \Delta g_2(e) = -1.$

Cell d has maximum gain $\Delta g_2 = 2$, $area(A) = 0$, balance criterion is violated.

Cell e has next maximum gain $\Delta g_2 = -1$, $area(A) = 9$, balance criterion is met.

Move cell e, updated partitions: $A_2 = \{d,e\}$, $B_2 = \{a,b,c\}$, with fixed cells $\{c,e\}$.

Pass 2, iteration $i = 3$

Gain values: $\Delta g_3(a) = 0, \Delta g_3(b) = -2, \Delta g_3(d) = 2.$

Cell d has maximum gain $\Delta g_3 = 2$, $area(A) = 5$, balance criterion is met.

Move cell d, updated partitions: $A_3 = \{e\}$, $B_3 = \{a,b,c,d\}$, with fixed cells $\{c,d,e\}$.

Pass 2, iteration $i = 4$

Gain values: $\Delta g_4(a) = -2, \Delta g_4(b) = -2.$

Cells a and b have maximum gain $\Delta g_4 = -2$.

Balance criterion after moving cell a: $area(A) = 7$.

Balance criterion after moving cell b: $area(A) = 9$.

Cell a meets the balance criterion better.

Move cell a, updated partitions: $A_4 = \{a,e\}$, $B_4 = \{b,c,d\}$, with fixed cells $\{a,c,d,e\}$.

Pass 2, iteration $i = 5$

Gain values: $\Delta g_5(b) = 1.$

Balance criterion after moving b: $area(A) = 11$.

Move cell b, updated partitions: $A_5 = \{a,b,e\}$, $B_5 = \{c,d\}$, with fixed cells $\{a,b,c,d,e\}$.

Find best move sequence $<c_1 \ldots c_m>$

$$G_1 = \Delta g_1 = 0$$
$$G_2 = \Delta g_1 + \Delta g_2 = -1$$
$$G_3 = \Delta g_1 + \Delta g_2 + \Delta g_3 = 1$$
$$G_4 = \Delta g_1 + \Delta g_2 + \Delta g_3 + \Delta g_4 = -1$$
$$G_5 = \Delta g_1 + \Delta g_2 + \Delta g_3 + \Delta g_4 + \Delta g_5 = 0$$

Maximum positive gain $G_m = 1$ occurs when $m = 3$.
Cells c, e and d are moved.

The result after Pass 2 is illustrated on the right.

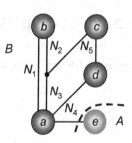

Exercise 4: System and Netlist Partitioning

One key difference is that traditional min-cut partitioning only accounts for minimizing net costs across k partitions. FPGA-based partitioning involves first determining the number of devices and then minimizing the total communication between devices as well as the device logic. For instance, traditional min-cut partitioning does not distinguish how many devices a p-pin net is split across. Furthermore, min-cut partitioning does not account for FPGA reconfigurability.

Exercise 5: Multilevel FM Partitioning

One major advantage is scalability. Traditional FM partitioning scales to ~200 nodes whereas multilevel FM can efficiently handle large-scale modern designs. The coarsening stage clusters nodes together, thereby reducing the number of nodes that FM interacts with. FM produces near-optimal solutions for netlists with fewer than 200 nodes, but solution quality deteriorates for larger netlists. In contrast, multilevel FM produces great solution quality without sacrificing large amounts of runtime.

Exercise 6: Clustering

Nodes that have either single connections to multiple other nodes or multiple connections to a single node are candidates for clustering. If a net *net* is contained within a single partition, then *net* does not contribute to the cut cost of the partition. If *net* spans multiple partitions, then one option is to place *net*'s cluster in the partition where *net*'s net weight is the greatest. Another option is to limit the size of the clusters such that the individual nodes of *net* are clustered within each partition.

Chapter 3: Chip Planning

Exercise 1: Slicing Trees and Constraint Graphs

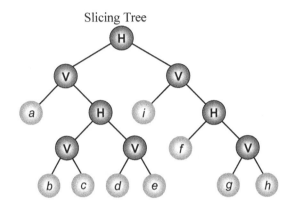

Slicing Tree

Vertical Constraint Graph (VCG)   Horizontal Constraint Graph (HCG)

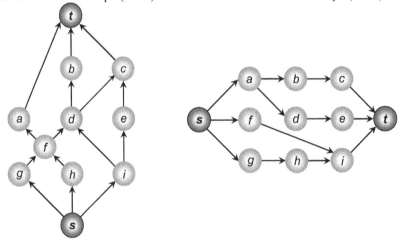

Exercise 2: Floorplan-Sizing Algorithm
(a) Shape functions for blocks a (left), b (center) and c (right).

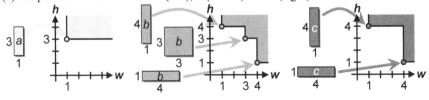

(b) Shape function of the floorplan.

Horizontal composition: Determine $h_{(a,b)}(w)$ of blocks a and b.

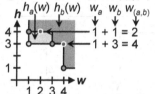

$1 + 1 = 2$
$1 + 3 = 4$

Vertical composition: Determine $h_{((a,b),c)}(w)$ and the minimum-area corner points. The minimum area of $((a,b),c)$ is 16, with dimensions either being $h_{((a,b),c)} = 8$ and $w_{((a,b),c)} = 2$, or $h_{((a,b),c)} = 8$ and $w_{((a,b),c)} = 4$.

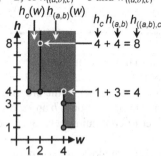
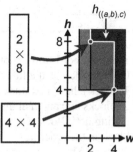

$4 + 4 = 8$
$1 + 3 = 4$

Construct floorplans by backtracing the shape functions of $h_c(w)$, $h_{(a,b)}(w)$, $h_a(w)$ and $h_b(w)$ to determine the dimensions of each block. The following shows the backtrace for the corner point (2,8) from $h_{((a,b),c)}(w)$.

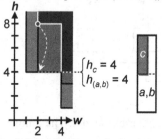
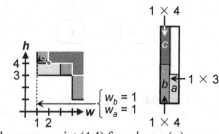

$\begin{cases} h_c = 4 \\ h_{(a,b)} = 4 \end{cases}$

$\begin{cases} w_b = 1 \\ w_a = 1 \end{cases}$

The following shows the backtrace for the corner point (4,4) from $h_{((a,b),c)}(w)$.

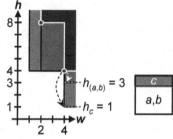
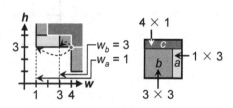

$h_{(a,b)} = 3$

$h_c = 1$

$w_b = 3$
$w_a = 1$

Exercise 3: Linear Ordering Algorithm

Iteration #	Block	New Nets	Terminating Nets	*gain*	Continuing Nets
0	*a*	N_1,N_2,N_3,N_4	--	-4	--
1	*b*	N_5,N_6	N_2	-1	--
	c	N_5	--	-1	N_3
	d	N_5,N_6	N_4	-1	N_3
	e	--	N_1	+1	--
2	*b*	N_5,N_6	N_2	-1	--
	c	N_5	--	-1	N_3
	d	N_5,N_6	N_4	-1	N_3
3	*b*	--	N_2,N_6	+2	N_5
	c	--	N_3	+1	N_5
4	*c*	--	N_3, N_5	+2	--

For each iteration, **bold font** denotes the block with the maximum gain.
Iteration 0: set block *a* as the first block in the ordering.
Iteration 1: block *e* has maximum gain. Set as the second block in the ordering.
Iteration 2: blocks *b*, *c* and *d* all have maximum gain of -1. Blocks *b* and *d* each have one terminating net. Block *d* has a higher number of continuing nets. Set block *d* as the third block in the ordering.
Iteration 3: block *b* has maximum gain. Set as the fourth block in the ordering.
Iteration 4: set block *c* as the fifth (last) block in the ordering.

The linear ordering that heuristically minimizes total net cost is <*a e d b c*>.

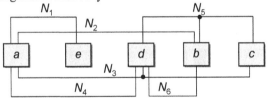

Exercise 4: Non-Slicing Floorplans
There are many possible non-slicing floorplans using four blocks *a-d*. The following is one such solution.

Chapter 4: Global and Detailed Placement

Exercise 1: Estimating Total Wirelength
(a) Tree representations of the five-pin net.

Minimum-Length Chain   Minimum Spanning Tree   Steiner Minimum Tree

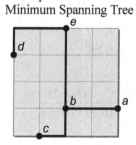 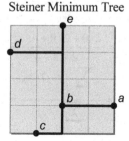

(b) $L(net) = \sum_{e \in net} w(e) \cdot d_M(e)$

$L_{Chain}(net) = w(d,e) \cdot d_M(d,e) + w(d,c) \cdot d_M(d,c) + w(c,b) \cdot d_M(c,b) + w(b,a) \cdot d_M(b,a)$
$\quad\quad\quad = 2 \cdot 3 + 2 \cdot 4 + 2 \cdot 2 + 2 \cdot 2 = 22$

$L_{MST}(net) = w(d,e) \cdot d_M(d,e) + w(e,b) \cdot d_M(e,b) + w(c,b) \cdot d_M(c,b) + w(b,a) \cdot d_M(b,a)$
$\quad\quad\quad = 2 \cdot 3 + 2 \cdot 3 + 2 \cdot 2 + 2 \cdot 2 = 20$

$L_{SMT}(net) = w(d,e,b) \cdot d_M(d,e,b) + w(c,b) \cdot d_M(c,b) + w(b,a) \cdot d_M(b,a)$
$\quad\quad\quad = 2 \cdot 5 + 2 \cdot 2 + 2 \cdot 2 = 18$

Exercise 2: Min-Cut Placement
After the initial partitioning (vertical cut cut_1):
Cells left of cut_1: $L = \{c,d,f,g\}$, cells right of cut_1: $R = \{a,b,e\}$. Cut cost: L-$R = 1$.

Possible solutions after the second partitioning (two horizontal cuts cut_{2L} and cut_{2R}):
Top Left: $TL = \{c,g\}$, Bottom Left: $BL = \{d,f\}$. Cut cost: TL-$BL = 2$.
Top Right: $TR = \{a\}$, Bottom Right: $BR = \{b,e\}$. Cut cost: TR-$BR = 1$.

After the third partitioning (four vertical cuts cut_{3TL}, cut_{3BL}, cut_{3TR} and cut_{3BR}), one possible result is

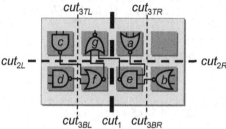

The maximum number of nets that cross an edge $\eta_P(e) = 2$, and the edge capacity $\sigma(e) = 2$. Therefore, since $\Phi(P) = 1$, the design is estimated to be fully routable.

Exercise 3: Force-Directed Placement

Solve for x_a^0 and x_b^0:

$$x_a^0 = \frac{\sum\limits_{c(a,j)\neq 0} c(a,j) \cdot x_j^0}{\sum\limits_{c(a,j)\neq 0} c(a,j)} = \frac{c(a,In1) \cdot x_{In1} + c(a,In2) \cdot x_{In2} + c(a,b) \cdot x_b^0}{c(a,In1) + c(a,In2) + c(a,b)}$$

$$= \frac{2 \cdot 0 + 2 \cdot 0 + 4 \cdot x_b^0}{2 + 2 + 4} = \frac{4x_b^0}{8} = 0.5x_b^0$$

$$x_b^0 = \frac{\sum\limits_{c(b,j)\neq 0} c(b,j) \cdot x_j^0}{\sum\limits_{c(b,j)\neq 0} c(b,j)} = \frac{c(b,a) \cdot x_a^0 + c(b,In2) \cdot x_{In2} + c(b,Out) \cdot x_{Out}}{c(b,a) + c(b,In2) + c(b,Out)}$$

$$= \frac{4 \cdot x_a^0 + 2 \cdot 0 + 2 \cdot 2}{4 + 2 + 2} = 0.5 + 0.5x_a^0$$

$$\left.\begin{array}{l} x_a^0 = 0.5x_b^0 \\ x_b^0 = 0.5 + 0.5x_a^0 \end{array}\right\} x_b^0 = 0.5 + 0.5(0.5x_b^0) = 2/3$$

$$x_a^0 = 0.5x_b^0 = 0.5(2/3) = 1/3$$

Rounded, $x_a^0 \approx 0$ and $x_b^0 \approx 1$.

Solve for y_a^0 and y_b^0:

$$y_a^0 = \frac{\sum\limits_{c(a,j)\neq 0} c(a,j) y_j^0}{\sum\limits_{c(a,j)\neq 0} c(a,j)} = \frac{c(a,In1) \cdot y_{In1} + c(a,In2) \cdot y_{In2} + c(a,b) \cdot y_b^0}{c(a,In1) + c(a,In2) + c(a,b)}$$

$$= \frac{2 \cdot 2 + 2 \cdot 0 + 4 \cdot y_b^0}{2 + 2 + 4} = 0.5 + 0.5y_b^0$$

$$y_b^0 = \frac{\sum\limits_{c(b,j)\neq 0} c(b,j) y_j^0}{\sum\limits_{c(b,j)\neq 0} c(b,j)} = \frac{c(b,a) \cdot y_a^0 + c(b,In2) \cdot y_{In2} + c(b,Out) \cdot y_{Out}}{c(b,a) + c(b,In2) + c(b,Out)}$$

$$= \frac{4 \cdot y_a^0 + 2 \cdot 0 + 2 \cdot 1}{4 + 2 + 2} = 0.25 + 0.5y_a^0$$

$$\left.\begin{array}{l} y_a^0 = 0.5 + 0.5y_b^0 \\ y_b^0 = 0.25 + 0.5y_a^0 \end{array}\right\} y_b^0 = 0.25 + 0.5(0.5 + 0.5y_b^0) = 2/3$$

$$y_a^0 = 0.5 + 0.5y_b^0 = 0.5 + 0.5(2/3) = 5/6$$

Rounded, $y_a^0 \approx 1$ and $y_b^0 \approx 1$.

The ZFT-positions for gates a and b are $(0,1)$ and $(1,1)$, respectively.

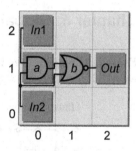

Exercise 4: Global and Detailed Placement

Global placement assigns cells or moveable blocks to bins of a coarse grid. In contrast, detailed placement moves cells within each grid bin such that cells do not mutually overlap (legalization). Typically, placement is split into two steps to ensure scalability. Legalization requires a large amount of runtime and cannot be applied after every iteration of global placement. A much faster method is to assign blocks to grid bins first and, when all blocks have been assigned, legalize afterward. Furthermore, detailed placement can be easily performed in parallel, as each grid bin can be processed independently. Typically, the placement process is split into two parts to ensure scalability.

Chapter 5: Global Routing

Exercise 1: Steiner Tree Routing

(a) All Hanan points and the minimum bounding box for the six-pin net.

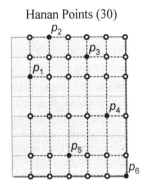
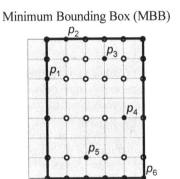

Hanan Points (30)   Minimum Bounding Box (MBB)

(b) Sequential Steiner tree heuristic.

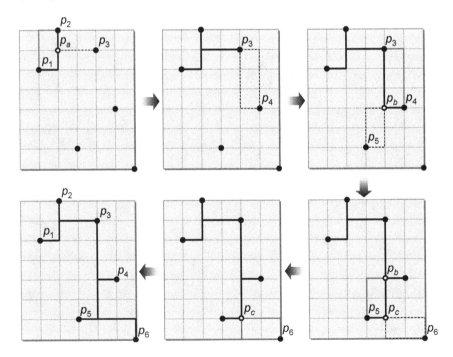

(c) Three Steiner points each with degree three.

(d) A three-pin net can have a maximum of one Steiner point.

Exercise 2: Global Routing in a Connectivity Graph
After routing net *A*.

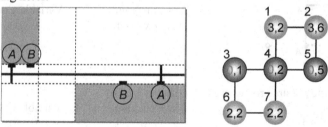

After routing net *B*.

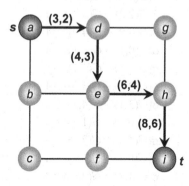

The given placement is routable.

Exercise 3: Dijkstra's Algorithm

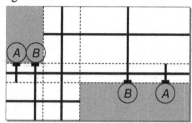

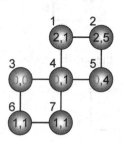

Group 2	Group 3
<a> [*b*] (2,2)	[*a*]
<a> [*d*] (3,2)	
~~*<b>* [*c*] (8,7)~~	*<a>* [*b*] (2,2)
~~*<b>* [*c*] (7,5)~~	
<d> [*e*] (4,3)	*<a>* [*d*] (3,2)
<d> [*g*] (8,5)	
<e> [*f*] (5,6)	*<d>* [*e*] (4,3)
<e> [*h*] (6,4)	
~~*<h>* [*g*] (9,5)~~	*<e>* [*h*] (6,4)
<h> [*i*] (8,6)	
<f> [*c*] (6,7)	*<e>* [*f*] (5,6)
~~*<f>* [*i*] (9,8)~~	
	<f> [*c*] (6,7)
	<d> [*g*] (8,5)
	<h> [*i*] (8,6)

Exercise 4: ILP-Based Global Routing

For net A, the possible routes are two L-shapes (A_1, A_2).

Net Constraints

$$x_{A_1} + x_{A_2} \leq 1$$

Variable Constraints

$$0 \leq x_{A_1} \leq 1$$
$$0 \leq x_{A_2} \leq 1$$

For net B, the possible routes are two L-shapes (B_1, B_2).

Net Constraints

$$x_{B_1} + x_{B_2} \leq 1$$

Variable Constraints

$$0 \leq x_{B_1} \leq 1$$
$$0 \leq x_{B_2} \leq 1$$

For net C, the possible routes are two L-shapes (C_1, C_2).

Net Constraints

$$x_{C_1} + x_{C_2} \leq 1$$

Variable Constraints

$$0 \leq x_{C_1} \leq 1$$
$$0 \leq x_{C_2} \leq 1$$

Horizontal Edge Capacity Constraints:

$G(0,0) \sim G(1,0)$:	x_{C_1}	$\leq$	$\sigma(G(0,0) \sim G(1,0)) = 1$
$G(1,0) \sim G(2,0)$:	x_{C_1}	$\leq$	$\sigma(G(1,0) \sim G(2,0)) = 1$
$G(2,0) \sim G(3,0)$:	x_{B_1}	$\leq$	$\sigma(G(2,0) \sim G(3,0)) = 1$
$G(3,0) \sim G(4,0)$:	x_{B_1}	$\leq$	$\sigma(G(3,0) \sim G(4,0)) = 1$
$G(0,1) \sim G(1,1)$:	x_{A_2}	$\leq$	$\sigma(G(0,1) \sim G(1,1)) = 1$
$G(1,1) \sim G(2,1)$:	x_{A_2}	$\leq$	$\sigma(G(1,1) \sim G(2,1)) = 1$
$G(2,1) \sim G(3,1)$:	x_{B_2}	$\leq$	$\sigma(G(2,1) \sim G(3,1)) = 1$
$G(3,1) \sim G(4,1)$:	x_{B_2}	$\leq$	$\sigma(G(3,1) \sim G(4,1)) = 1$
$G(0,2) \sim G(1,2)$:	x_{C_2}	$\leq$	$\sigma(G(0,2) \sim G(1,2)) = 1$
$G(1,2) \sim G(2,2)$:	x_{C_2}	$\leq$	$\sigma(G(1,2) \sim G(2,2)) = 1$
$G(0,3) \sim G(1,3)$:	x_{A_1}	$\leq$	$\sigma(G(0,3) \sim G(1,3)) = 1$
$G(1,3) \sim G(2,3)$:	x_{A_1}	$\leq$	$\sigma(G(1,3) \sim G(2,3)) = 1$

Vertical Edge Capacity Constraints:

$G(0,0) \sim G(0,1)$:	x_{C_2}	$\leq$	$\sigma(G(0,0) \sim G(0,1)) = 1$
$G(2,0) \sim G(2,1)$:	$x_{B_2} + x_{C_1}$	$\leq$	$\sigma(G(2,0) \sim G(2,1)) = 1$
$G(4,0) \sim G(4,1)$:	x_{B_1}	$\leq$	$\sigma(G(4,0) \sim G(4,1)) = 1$
$G(0,1) \sim G(0,2)$:	$x_{A_2} + x_{C_2}$	$\leq$	$\sigma(G(0,1) \sim G(0,2)) = 1$
$G(2,1) \sim G(2,2)$:	$x_{A_1} + x_{C_1}$	$\leq$	$\sigma(G(2,1) \sim G(2,2)) = 1$
$G(0,2) \sim G(0,3)$:	x_{A_2}	$\leq$	$\sigma(G(0,2) \sim G(0,3)) = 1$
$G(2,2) \sim G(2,3)$:	x_{A_1}	$\leq$	$\sigma(G(2,2) \sim G(2,3)) = 1$

Objective Function: Maximize $x_{A_1} + x_{A_2} + x_{B_1} + x_{B_2} + x_{C_1} + x_{C_2}$

The solution is routable with routes A_2, B_1, and C_1 (right).

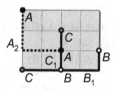

Exercise 5: Shortest Path with A* Search
Remove the top right obstacle yields the following A* search progression.

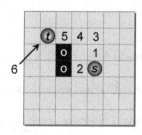

Exercise 6: Rip-Up and Reroute
Estimated memory usage is $O(m^2 \cdot n)$. The number of net segments per net is $O(m^2)$. Since the number of nets is n, the tight upper bound of memory usage is $O(m^2 \cdot n)$.

Chapter 6: Detailed Routing

Exercise 1: Left-Edge Algorithm

(a) Sets $S(col)$:   Maximal $S(col)$:

    $S(a) = \{A,B\}$

    $S(b) = \{A,B,C\}$ $S(b) = \{A,B,C\}$

    $S(c) = \{A,C,D\}$ $S(c) = \{A,C,D\}$ or $S(d) = \{A,C,D\}$

    $S(d) = \{A,C,D\}$

    $S(e) = \{C,D,E\}$ $S(e) = \{C,D,E\}$

    $S(f) = \{D,E,F\}$ $S(f) = \{D,E,F\}$

    $S(g) = \{E,F\}$

    $S(h) = \{F\}$

Minimum number of required tracks = $|S(b)| = |S(c)| = |S(d)| = |S(e)| = |S(f)| = 3$.

(b) Horizontal constraint graph (HCG) and vertical constraint graph (VCG).

Horizontal Constraint Graph (HCG)

Vertical Constraint Graph (VCG)

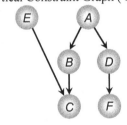

(c) Routed channel.

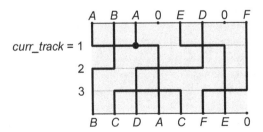

Exercise 2: Dogleg Left-Edge Algorithm

(a) Vertical constraint graph (VCG) without net splitting.

(b) After splitting nets A, C and D: $\{A_1,A_2,A_3,B,C_1,C_2,D_1,D_2,E\}$.

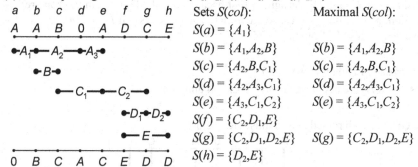

	Sets $S(col)$:	Maximal $S(col)$:
	$S(a) = \{A_1\}$	
	$S(b) = \{A_1,A_2,B\}$	$S(b) = \{A_1,A_2,B\}$
	$S(c) = \{A_2,B,C_1\}$	$S(c) = \{A_2,B,C_1\}$
	$S(d) = \{A_2,A_3,C_1\}$	$S(d) = \{A_2,A_3,C_1\}$
	$S(e) = \{A_3,C_1,C_2\}$	$S(e) = \{A_3,C_1,C_2\}$
	$S(f) = \{C_2,D_1,E\}$	
	$S(g) = \{C_2,D_1,D_2,E\}$	$S(g) = \{C_2,D_1,D_2,E\}$
	$S(h) = \{D_2,E\}$	

(c) Vertical constraint graph (VCG) after net splitting.

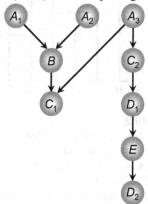

(d) Without net splitting, the instance is not routable because of the cycle D-E-D in the VCG from (a). With net splitting, the instance is routable. The minimum number of tracks needed = $|S(c)| = |S(g)| = 3$.

(e) Track assignment:

curr_track = 1

Consider nets A_1, A_2 and A_3. Assign A_1 first because it is the leftmost in the zone representation. Nets A_2 and A_3 do not cause a conflict. Therefore, assign A_2 and A_3 to *curr_track*. Remove nets A_1, A_2 and A_3 from the VCG.

curr_track = 2

Consider nets B and C_2. Assign B first because it is the leftmost in the zone representation. Net C_2 does not cause a conflict. Therefore, assign C_2 to *curr_track*. Remove nets B and C_2 from the VCG.

curr_track = 3

Consider nets C_1 and D_1. Assign C_1 first because it is the leftmost in the zone representation. Net D_1 does not cause a conflict. Therefore, assign D_1 to *curr_track*. Remove nets C_1 and D_1 from the VCG.

curr_track = 4 and *curr_track* = 5

Nets E and D_2 are assigned to *curr_track* = 4 and *curr_track* = 5.

The channel with routed nets is illustrated below.

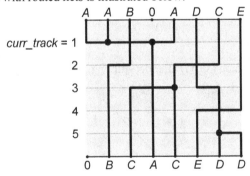

Exercise 3: Switchbox Routing

The switchbox has six columns a-f from left to right, and six tracks 1-6 from bottom to top. The following are the steps carried out at each column.

Column a: Assign net F to track 1. Assign net B to track 6. Extend nets F (track 1), G (track 2), A (track 3) and B (track 6).

Column b: Connect the top pin A and bottom pin A to net A on track 3. Extend nets F (track 1), G (track 2) and B (track 6).

Column *c*: Connect the bottom pin *F* to net *F* on track 1. Connect the top pin *C* to track 3. Extend nets *G* (track 2), *C* (track 3) and *B* (track 6).

Column *d*: Connect the bottom pin *G* to net *G* on track 2. Connect the top pin *E* to track 4. Extend nets *G* (track 2), *C* (track 3), *E* (track 4) and *B* (track 6).

Column *e*: Connect the bottom pin *D* with track 1. Connect the top pin *B* to net *B* on track 6. Assign net *G* to track 5. Extend nets *D* (track 1), *C* (track 3), *E* (track 4) and *G* (track 5).

Column *f*: Connect the top pin *D* to net *D* on track 1. Extend the nets on tracks 2, 3, 4 and 5 to their corresponding pins.

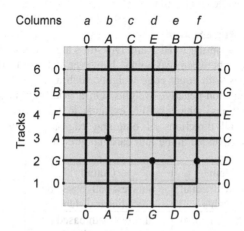

Exercise 4: Manufacturing Defects

In a congested region, connections are more likely to detour, which increases the usage of vias. Therefore, vias are more likely to occur. Likewise, since wires are packed more closely, shorts are more likely in congested regions. Opens are also more likely because detoured connections are longer. Antenna effects are, *a priori*, less likely in congested regions because fewer connections use long straightline segments on low metal layers. However, routing congestion makes it more difficult to fix antenna violations.

Exercise 5: Modern Challenges in Detailed Routing

See reference [6.17] - G. Xu, L.-D. Huang, D. Pan and M. Wong, "Redundant-Via Enhanced Maze Routing for Yield Improvement", *Proc. Asia and South Pacific Design Autom. Conf.*, 2005, pp. 1148-1151.

Exercise 6: Non-Tree Routing

One advantage of non-tree routing is that the redundant wires mitigate the impact of opens. However, redundant routing increases the total wirelength of the design. Excessive wirelength, especially in congested regions, pushes wires closer to each other, increasing the incidence of shorts. More information on non-tree routing can be found in [6.12] - A. Kahng, B. Liu and I. Măndoiu, "Non-Tree Routing for Reliability and Yield Improvement", *Proc. Intl. Conf. on CAD*, 2002, pp. 260-266.

Chapter 7: Specialized Routing

Exercise 1: Net Ordering
Without history costs, there is no indication how much demand a given edge or track has over time. In contrast, congestion only shows how often a track is used in the recent past, e.g., the previous or current iteration. That is, without any previous knowledge of rip-up and reroute iterations, a router has very limited knowledge about which tracks are frequently used. Therefore, net ordering is the primary method in which to alleviate congestion.

Exercise 2: Octilinear Maze Search
Octilinear maze search is similar to BFS and Dijkstra's algorithm in the sense that it can find the shortest path between two points. However, octilinear maze search and BFS only run on graphs with equal edge weights, such as a grid, while Dijkstra's can run on graphs with non-negative weights. On a grid, octilinear maze search expands in eight directions, whereas BFS expands in four directions. Dijkstra's algorithm expands in all directions of outgoing edges. In general, on a grid, Dijkstra's algorithm is expanded in the four cardinal directions.

Algorithm	Input	Output	Runtime
Octilinear maze search	graph with equal weights	Shortest path based on octilinear distance	$O(|V| + |E|)$
BFS	graph with equal weights	Shortest path based on Manhattan distance	$O(|V| + |E|)$
Dijkstra's algorithm	graph with non-negative weights	Shortest path based on edge weights	$O(|V| \log |E|)$

Exercise 3: Method of Means and Medians (MMM)
(a) Clock tree constructed by MMM.

$$x_c(s_1 - s_8) = \frac{2+4+2+4+8+8+14+14}{8}$$

$$= \frac{56}{8} = 7$$

$$y_c(s_1 - s_8) = \frac{2+2+12+12+8+6+8+6}{8}$$

$$= \frac{56}{8} = 7$$

Route s_0 to u_1 (7,7). After dividing by the median, s_1-s_4 are in one set, and s_5-s_8 are in another set.

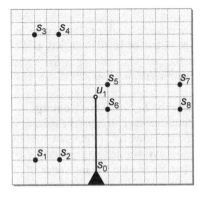

$$x_c(s_1 - s_4) = \frac{2+4+2+4}{4} = \frac{12}{4} = 3$$

$$y_c(s_1 - s_4) = \frac{2+2+12+12}{4} = \frac{28}{4} = 7$$

Route u_1 to u_2 (3,7). After dividing by the median, s_1 and s_2 are in one set, and s_3 and s_4 are in another set.

$$x_c(s_5 - s_8) = \frac{8+8+14+14}{4} = \frac{44}{4} = 11$$

$$y_c(s_5 - s_8) = \frac{8+6+8+6}{4} = \frac{28}{4} = 7$$

Route u_1 to u_3 (11,7). After dividing by the median, s_5 and s_7 are in one set, and s_6 and s_8 are in another set.

$$x_c(s_1 - s_2) = \frac{2+4}{2} = \frac{6}{2} = 3$$

$$y_c(s_1 - s_2) = \frac{2+2}{2} = \frac{4}{2} = 2$$

$$x_c(s_3 - s_4) = \frac{2+4}{2} = \frac{6}{2} = 3$$

$$y_c(s_3 - s_4) = \frac{12+12}{2} = \frac{24}{2} = 12$$

Route u_4 (3,2) and u_5 (3,12) to u_2.

$$x_c(s_5, s_7) = \frac{8+14}{2} = \frac{22}{2} = 11$$

$$y_c(s_5, s_7) = \frac{8+8}{2} = \frac{16}{2} = 8$$

$$x_c(s_6, s_8) = \frac{8+14}{2} = \frac{22}{2} = 11$$

$$y_c(s_6, s_8) = \frac{6+6}{2} = \frac{12}{2} = 6$$

Route u_6 (11,8) and u_7 (11,6) to u_3.

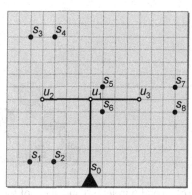

Route sinks s_1 and s_2 to u_4.

Route sinks s_3 and s_4 to u_5.

Route sinks s_5 and s_7 to u_6.

Route sinks s_6 and s_8 to u_7.

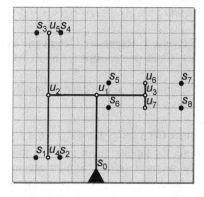

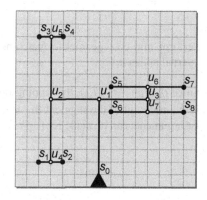

(b) Topology of the clock tree generated by MMM

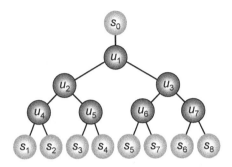

(c) Total wirelength and skew of the clock tree T generated by MMM.
$L(T)$ = total number of grid edges = 42.

$t_{LD}(s_0,s_1) = t_{LD}(s_0,s_2) = t_{LD}(s_0,s_3) = t_{LD}(s_0,s_4) = t_{LD}(s_0,s_1\text{-}s_4) = 16$
$t_{LD}(s_0,s_5) = t_{LD}(s_0,s_6) = t_{LD}(s_0,s_7) = t_{LD}(s_0,s_8) = t_{LD}(s_0,s_5\text{-}s_8) = 14$
$skew(T) = |t_{LD}(s_0,s_1\text{-}s_4) - t_{LD}(s_0,s_5\text{-}s_8)| = |16 - 14| = 2.$

Exercise 4: Recursive Geometric Matching (RGM) Algorithm
(a) Clock tree constructed by RGM.

Perform min-cost geometric matching over sinks s_1-s_8.

Since $C(s_1) = C(s_2)$, $t_{LD}(T_{s_1}) = t_{LD}(T_{s_2})$. Therefore, the tapping point u_1 is the midpoint of the segment $(s_1 \sim s_2)$. Similar arguments can be made for $(s_3 \sim s_4)$, $(s_5 \sim s_6)$ and $(s_7 \sim s_8)$.

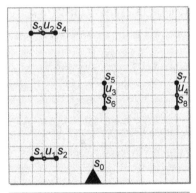

Perform min-cost geometric matching over internal nodes u_1-u_4.

Since $t_{LD}(T_{u_1}) = t_{LD}(T_{u_2})$, the tapping point u_5 is the midpoint of the segment $(u_1 \sim u_2)$. Similar arguments can be made for $(u_3 \sim u_4)$.

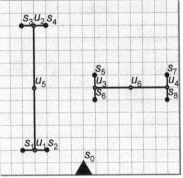

Perform min-cost, geometric matching over internal nodes u_5 and u_6. For tapping point u_7 on $(u_5 \sim u_6)$,

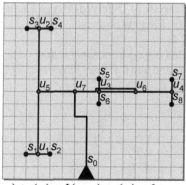

$$t_{LD}(T_{u_5}) + t_{LD}(u_5, u_7) = t_{LD}(T_{u_6}) + t_{LD}(u_7, u_6)$$

$t_{LD}(u_5, u_7) = L(u_5, u_7)$, and $t_{LD}(u_7, u_6) = L(u_7, u_6)$.
Since $L(u_5, u_7) + L(u_7, u_6) = L(u_5, u_6) = 8$,

$$t_{LD}(u_7, u_6) = L(u_5, u_6) - L(u_5, u_7)$$

$t_{LD}(T_{u_5}) = L(u_5, s_1) + t(s_1) = L(u_5, s_2) + t(s_2) = L(u_5, s_3) + t(s_3) = L(u_5, s_4) + t(s_4) = 6.$
$t_{LD}(T_{u_6}) = L(u_6, s_5) + t(s_5) = L(u_6, s_6) + t(s_6) = L(u_6, s_7) + t(s_7) = L(u_6, s_8) + t(s_8) = 4.$
Combing all above equations, $L(u_5, u_7) = 3$ and $L(u_7, u_6) = 5$. Route s_0 to u_7.

(b) Topology of the clock tree generated by RGM.

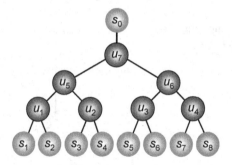

(c) Total wirelength and skew of the clock tree T generated by MMM.
$L(T)$ = total number of grid edges = 39.

$$t_{LD}(s_0, s_1) = t_{LD}(s_0, s_2) = t_{LD}(s_0, s_3) = t_{LD}(s_0, s_4) = t_{LD}(s_0, s_5)$$
$$= t_{LD}(s_0, s_6) = t_{LD}(s_0, s_7) = t_{LD}(s_0, s_8) = 16.$$

Since the delay from s_0 to all sinks is the same, $skew(T) = 0$.

Exercise 5: Deferred-Merge Embedding (DME) Algorithm

(a) Merging segments.

Find $ms(u_2)$ from sinks s_1 and s_3, and $ms(u_3)$ from sinks s_2 and s_4.

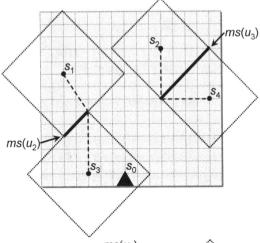

Find $ms(u_1)$ from merging segments $ms(u_2)$ and $ms(u_3)$.

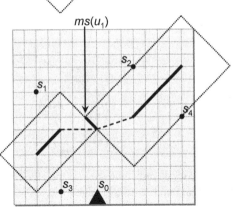

(b) Embedded clock tree.

Connect s_0 to u_1. Based on $trr(u_1)$, connect $ms(u_2)$ and $ms(u_3)$ to u_1.

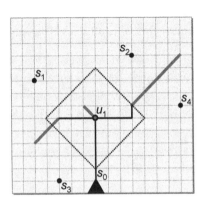

Based on $trr(u_2)$, route u_2 to sinks s_1 and s_3. Based on $trr(u_3)$, route u_3 to sinks s_2 and s_4.

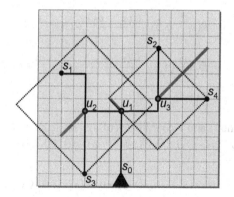

Exercise 6: Exact Zero Skew
(a) Location of internal nodes u_1, u_2 and u_3.

Find x- and y-coordinates for u_2 – on segment between s_1 and s_3.

$t_{ED}(s_1) = 0$, $t_{ED}(s_3) = 0$, $C(s_1) = 0.1$, $C(s_3) = 0.3$, $\alpha = 0.1$, $\beta = 0.01$

$L(s_1,s_3) = |x_{s_1} - x_{s_3}| + |y_{s_1} - y_{s_3}| = |2 - 4| + |9 - 1| = 10$

$$z_{s_1 \sim u_2} = \frac{(t_{ED}(s_3) - t_{ED}(s_1)) + \alpha \cdot L(s_1,s_3) \cdot \left(C(s_3) + \dfrac{\beta \cdot L(s_1,s_3)}{2} \right)}{\alpha \cdot L(s_1,s_3) \cdot (\beta \cdot L(s_1,s_3) + C(s_1) + C(s_3))}$$

$$= \frac{(0 - 0) + 0.1 \cdot 10 \cdot \left(0.3 + \dfrac{0.01 \cdot 10}{2} \right)}{0.1 \cdot 10 \cdot (0.01 \cdot 10 + 0.1 + 0.3)} = \frac{0.35}{0.5} = 0.7$$

$z_{s_1 \sim u_2} \cdot L(s_1,s_3) = 0.7 \cdot 10 = 7$, x- and y-coordinates for $u_2 = (4,4)$.

Find the capacitance $C(u_2)$.

$C(s_1) = 0.1$, $C(s_3) = 0.3$, $\beta = 0.01$, $L(s_1,s_3) = 10$

$C(u_2) = C(s_1) + C(s_3) + \beta \cdot L(s_1,s_3) = 0.1 + 0.3 + 0.01 \cdot 10 = 0.5$

Find the delay $t_{ED}(T_{u_2})$.

$t_{ED}(s_1) = 0$, $t_{ED}(s_3) = 0$, $z_{s_1 \sim u_2} = 0.7$, $z_{u_2 \sim s_3} = 1 - z_{s_1 \sim u_2} = 1 - 0.7 = 0.3$

$R(s_1 \sim u_2) = \alpha \cdot z_{s_1 \sim u_2} \cdot L(s_1,s_3) = 0.1 \cdot 0.7 \cdot 10 = 0.7$
$C(s_1 \sim u_2) = \beta \cdot z_{s_1 \sim u_2} \cdot L(s_1,s_3) = 0.01 \cdot 0.7 \cdot 10 = 0.07$

$R(u_2 \sim s_3) = \alpha \cdot z_{u_2 \sim s_3} \cdot L(s_1,s_3) = 0.1 \cdot 0.3 \cdot 10 = 0.3$
$C(u_2 \sim s_3) = \beta \cdot z_{u_2 \sim s_3} \cdot L(s_1,s_3) = 0.01 \cdot 0.3 \cdot 10 = 0.03$

$$t_{ED}(u_2) = R(s_1 \sim u_2) \cdot \left(\frac{C(s_1 \sim u_2)}{2} + C(s_1) \right) + t_{ED}(s_1) = 0.7 \cdot \left(\frac{0.07}{2} + 0.1 \right) + 0$$

$$= R(u_2 \sim s_3) \cdot \left(\frac{C(u_2 \sim s_3)}{2} + C(s_3) \right) + t_{ED}(s_3) = 0.3 \cdot \left(\frac{0.03}{2} + 0.3 \right) + 0$$

$$= 0.0945$$

Find the x- and y-coordinates of u_3 – on segment between s_2 and s_4.
$t_{ED}(s_2) = 0$, $t_{ED}(s_4) = 0$, $C(s_2) = 0.2$, $C(s_4) = 0.2$, $\alpha = 0.1$, $\beta = 0.01$, $L(s_2,s_4) = 8$

$$z_{s_2 \sim u_3} = \frac{(t_{ED}(s_4) - t_{ED}(s_2)) + \alpha \cdot L(s_2, s_4) \cdot \left(C(s_4) + \dfrac{\beta \cdot L(s_2, s_4)}{2} \right)}{\alpha \cdot L(s_2, s_4) \cdot (\beta \cdot L(s_2, s_4) + C(s_2) + C(s_4))}$$

$$= \frac{(0 - 0) + 0.1 \cdot 8 \cdot \left(0.2 + \dfrac{0.01 \cdot 8}{2} \right)}{0.1 \cdot 8 \cdot (0.01 \cdot 8 + 0.2 + 0.2)} = \frac{0.192}{0.384} = 0.5$$

$z_{s_2 \sim u_3} \cdot L(s_2,s_4) = 0.5 \cdot 8 = 4$, x- and y-coordinates for $u_3 = (10,7)$.

Find the capacitance $C(u_3)$.
$C(s_2) = 0.2$, $C(s_4) = 0.2$, $\beta = 0.01$, $L(s_2,s_4) = 8$

$$C(u_3) = C(s_2) + C(s_4) + \beta \cdot L(s_2,s_4) = 0.2 + 0.2 + 0.01 \cdot 8 = 0.48$$

Find the delay $t_{ED}(u_3)$.
$t_{ED}(s_2) = 0$, $t_{ED}(s_4) = 0$, $z_{s_2 \sim u_3} = 0.5$, $z_{u_3 \sim s_4} = 1 - z_{s_2 \sim u_3} = 1 - 0.5 = 0.5$

$$R(s_2 \sim u_3) = \alpha \cdot z_{s_2 \sim u_3} \cdot L(s_2,s_4) = 0.1 \cdot 0.5 \cdot 8 = 0.4$$
$$C(s_2 \sim u_3) = \beta \cdot z_{s_2 \sim u_3} \cdot L(s_2,s_4) = 0.01 \cdot 0.5 \cdot 8 = 0.04$$

$$R(u_3 \sim s_4) = \alpha \cdot z_{u_3 \sim s_4} \cdot L(s_2,s_4) = 0.1 \cdot 0.5 \cdot 8 = 0.4$$
$$C(u_3 \sim s_4) = \beta \cdot z_{u_3 \sim s_4} \cdot L(s_2,s_4) = 0.01 \cdot 0.5 \cdot 8 = 0.04$$

$$t_{ED}(u_3) = R(s_2 \sim u_3) \cdot \left(\frac{C(s_2 \sim u_3)}{2} + C(s_2) \right) + t_{ED}(s_2) = 0.4 \cdot \left(\frac{0.04}{2} + 0.2 \right) + 0$$

$$= R(u_3 \sim s_4) \cdot \left(\frac{C(u_3 \sim s_4)}{2} + C(s_4) \right) + t_{ED}(s_4) = 0.4 \cdot \left(\frac{0.04}{2} + 0.2 \right) + 0$$

$$= 0.088$$

Find the x- and y-coordinates of u_1 – on segment between u_2 and u_3.
$t_{ED}(u_2) = 0.0945$, $t_{ED}(u_3) = 0.088$, $C(u_2) = 0.5$, $C(u_3) = 0.48$, $\alpha = 0.1$, $\beta = 0.01$
$L(u_2,u_3) = |x_{u_2} - x_{u_3}| + |y_{u_2} - y_{u_3}| = |4 - 10| + |4 - 7| = 9$

$$z_{u_2 \sim u_1} = \frac{(t_{ED}(u_3) - t_{ED}(u_2)) + \alpha \cdot L(u_2, u_3) \cdot \left(C(u_3) + \dfrac{\beta \cdot L(u_2, u_3)}{2}\right)}{\alpha \cdot L(u_2, u_3) \cdot (\beta \cdot L(u_2, u_3) + C(u_2) + C(u_3))}$$

$$= \frac{(0.088 - 0.0945) + 0.1 \cdot 9 \cdot \left(0.48 + \dfrac{0.01 \cdot 9}{2}\right)}{0.1 \cdot 9 \cdot (0.01 \cdot 9 + 0.5 + 0.48)} = \frac{0.466}{0.963} \approx 0.484$$

$z_{u_2 \sim u_1} \cdot L(u_2, u_3) = 0.484 \cdot 9 \approx 4.36$, x- and y-coordinates for $v_1 = (8.36, 4)$.

(b) Elmore delay from u_1 to every sink s_1-s_4.
$t_{ED}(u_2) = 0.0945$, $t_{ED}(u_3) = 0.088$, $C(u_2) = 0.5$, $C(u_3) = 0.48$,
$z_{u_2 \sim u_1} = z_{u_1 \sim u_2} \approx 0.484$, $z_{u_1 \sim u_3} = 1 - z_{u_2 \sim u_1} \approx 0.516$

$R(u_1 \sim u_2) = \alpha \cdot z_{u_1 \sim u_2} \cdot L(u_2, u_3) = 0.1 \cdot 0.484 \cdot 9 = 0.4356$
$C(u_1 \sim u_2) = \beta \cdot z_{u_1 \sim u_2} \cdot L(u_2, u_3) = 0.01 \cdot 0.484 \cdot 9 = 0.04356$

$R(u_1 \sim u_3) = \alpha \cdot z_{u_1 \sim u_3} \cdot L(u_2, u_3) = 0.1 \cdot 0.516 \cdot 9 = 0.4644$
$C(u_1 \sim u_3) = \beta \cdot z_{u_1 \sim u_3} \cdot L(u_2, u_3) = 0.01 \cdot 0.516 \cdot 9 = 0.04644$

$$t_{ED}(u_1, s_1\text{-}s_4) = t_{ED}(u_1, s_1) = t_{ED}(u_1, s_3)$$

$$= R(u_1 \sim u_2) \cdot \left(\frac{C(u_1 \sim u_2)}{2} + C(u_2)\right) + t_{ED}(u_2)$$

$$= 0.4356 \cdot \left(\frac{0.04356}{2} + 0.5\right) + 0.0945$$

$$= t_{ED}(u_1, s_2) = t_{ED}(u_1, s_4)$$

$$= R(u_1 \sim u_3) \cdot \left(\frac{C(u_1 \sim u_3)}{2} + C(u_3)\right) + t_{ED}(u_3)$$

$$= 0.4644 \cdot \left(\frac{0.04644}{2} + 0.48\right) + 0.088$$

$$\approx 0.322$$

Exercise 7: Bounded-Skew DME
Feasible regions are marked as follows.

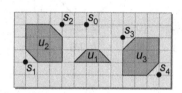

Chapter 8: Timing Closure

Exercise 1: Static Timing Analysis
(a) Timing graph.

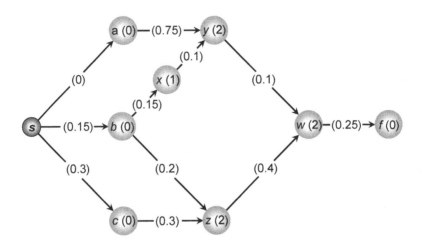

(b) Actual arrival times.

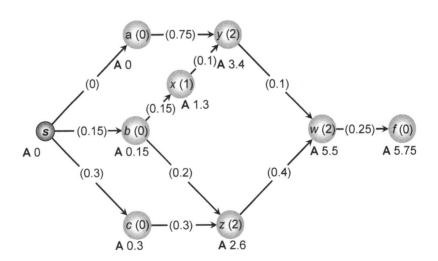

(c) Required arrival times.

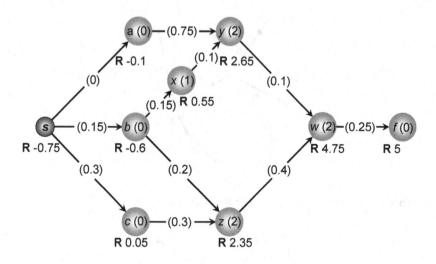

(d) Slacks.

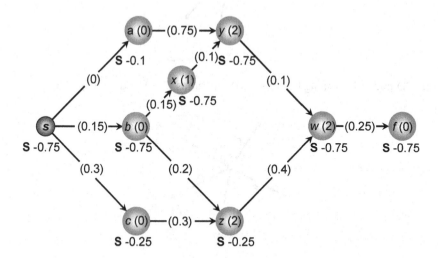

Exercise 2: Timing-Driven Routing

(a)   Minimum spanning tree with radius(T) = 13 and cost(T) = 30.

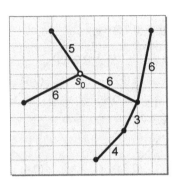

(b)   Shortest-paths tree with radius(T) = 8 and cost(T) = 39.

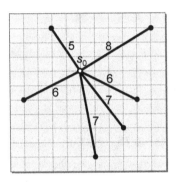

(c)   PD-tradeoff spanning tree (γ = 0.5) with radius(T) = 9 and cost(T) = 35.

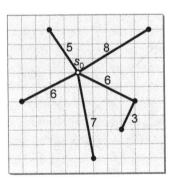

Exercise 3: Buffer Insertion for Timing Improvement
With Fig. 8.12, the delay of each gate can be calculated with its load capacitance. Buffer y always has a load capacitance of 2.5 fF.

Buffer y with size A: v_B has load capacitance = 2.5 fF, which results in $t(v_B) = 30$ ps. $AAT(c) = t(v_B) + t(y_A) = 30 + 35 = 65$ ps.

Buffer y with size B, v_B has load capacitance = 3 fF, which results in $t(v_B) = 33$ ps. $AAT(c) = t(v_B) + t(y_B) = 33 + 30 = 63$ ps.

Buffer y with size C: v_B has load capacitance = 4 fF, which results in $t(v_B) = 39$ ps. $AAT(c) = t(v_B) + t(y_C) = 39 + 27 = 66$ ps.

The best size for buffer y is B.

Exercise 4: Timing Optimization
1. Delay budgeting: assigning upper bounds on timing or length for nets. These limits restrict the maximum amount of time a signal travels along critical nets. However, if too many nets are constrained, this can lead to wirelength degradation or highly-congested regions.
2. Physical synthesis, such as gate sizing and cloning. Sizing up gates can improve the delay on specific paths at the cost of increased area and power. Cloning can mitigate long interconnect delays by duplicating gates or signals at locations closer to the desired location.

Exercise 5: Cloning vs. Buffering
Cloning is more advantageous than buffering when the same timing-critical signal is needed in multiple locations that are relatively far apart. The signal can just be reproduced locally. This can save on area and routing resources. Buffering can be more advantageous than cloning since buffers do not increase the upstream capacitance of the gate, which is helpful in terms of circuit delay and power.

Exercise 6: Physical Synthesis
Buffer removal (vs. buffer insertion): If the placement or routing of the buffered net changes, some buffers may no longer be necessary to meet timing constraints. Alternatively, buffers can be removed if the net is not timing-critical or has positive slack.

Gate downsizing (vs. gate upsizing): If the path that goes through the gate can be slowed down without slack violations, then the gate can be downsized.

Merging (vs. cloning): If the netlist, placement or routing of the design changes, some nodes could be removed due to redundancy.

For all three transforms, increasing the area can cause illegality (overlap) in the placement. Therefore, the reverse transforms can be necessary to meet area constraints or relax timing for non-critical paths.

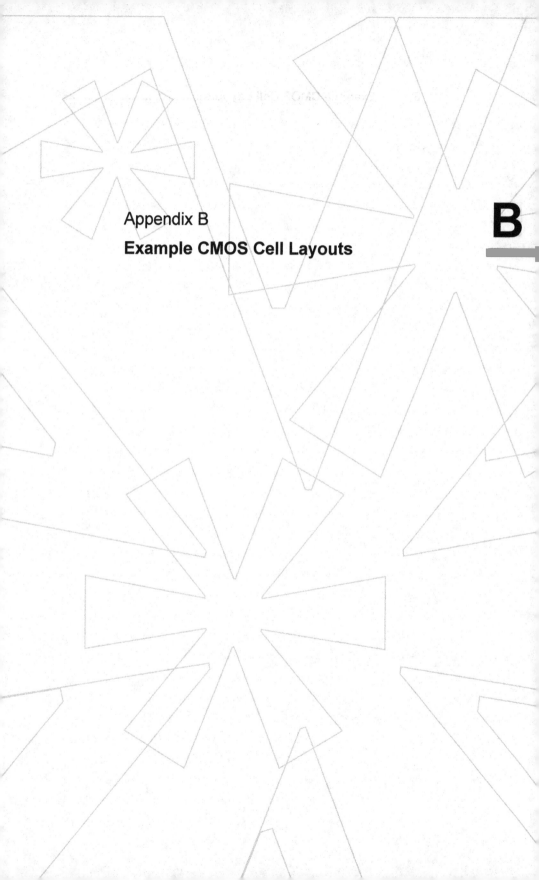

Appendix B

Example CMOS Cell Layouts

B

B

B Example CMOS Cell Layouts

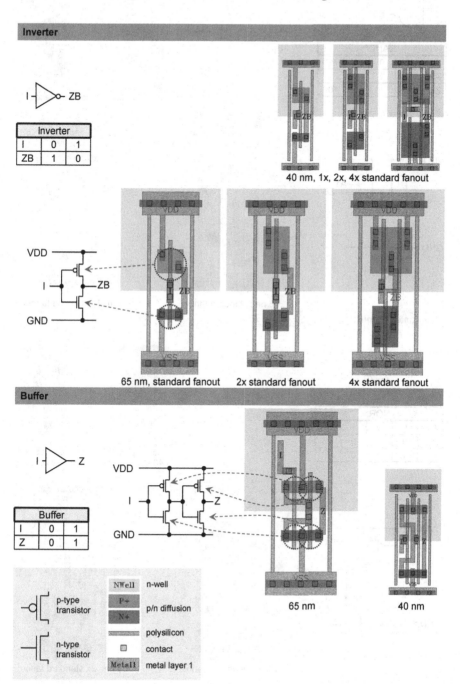

Inverter

I —▷o— ZB

Inverter		
I	0	1
ZB	1	0

40 nm, 1x, 2x, 4x standard fanout

VDD
I — ZB
GND

65 nm, standard fanout   2x standard fanout   4x standard fanout

Buffer

I —▷— Z

Buffer		
I	0	1
Z	0	1

VDD
I — Z
GND

65 nm   40 nm

| | p-type transistor |
| | n-type transistor |

NWell	n-well
P+	p/n diffusion
N+	
	polysilicon
□	contact
Metal1	metal layer 1

NAND gate (2-input)

	NAND			
A1	0	0	1	1
A2	0	1	0	1
ZB	1	1	1	0

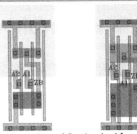

40 nm, 1x and 2x standard fanout

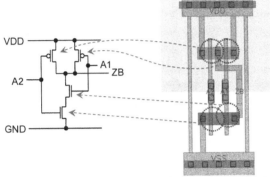

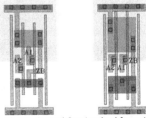

65 nm, standard fanout   65 nm, 2x standard fanout

NOR gate (2-input)

	NOR			
A1	0	0	1	1
A2	0	1	0	1
ZB	1	0	0	0

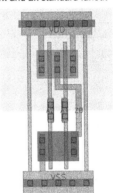

40 nm, 1x and 2x standard fanout

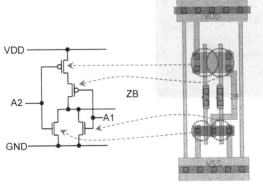

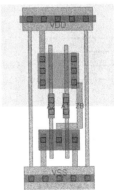

65 nm, standard fanout   65 nm, 2x standard fanout

AND-OR-Invert (AOI) gate (2-1)

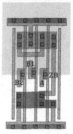
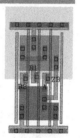

AOI								
A	0	0	0	0	1	1	1	1
B1	0	0	1	1	0	0	1	1
B2	0	1	0	1	0	1	0	1
ZB	1	1	1	0	0	0	0	0

40 nm, 1x and 2x standard fanout

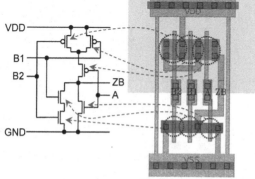
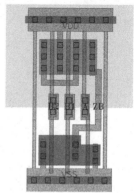

65 nm, standard fanout   **65 nm, 2x standard fanout**

Index

D